Gilmar Paulinho Triches

Eucalyptus and Cattle in Integrated Systems: behavior and damage

AF515284

Gilmar Paulinho Triches

Eucalyptus and Cattle in Integrated Systems: behavior and damage

Initial Behavior of Eucalyptus benthamii Maiden et Cambage in Integrated Agricultural Production Systems

ScienciaScripts

Imprint
Any brand names and product names mentioned in this book are subject to trademark, brand or patent protection and are trademarks or registered trademarks of their respective holders. The use of brand names, product names, common names, trade names, product descriptions etc. even without a particular marking in this work is in no way to be construed to mean that such names may be regarded as unrestricted in respect of trademark and brand protection legislation and could thus be used by anyone.

Cover image: www.ingimage.com

This book is a translation from the original published under ISBN 978-613-9-61538-4.

Publisher:
Sciencia Scripts
is a trademark of
Dodo Books Indian Ocean Ltd. and OmniScriptum S.R.L publishing group

120 High Road, East Finchley, London, N2 9ED, United Kingdom
Str. Armeneasca 28/1, office 1, Chisinau MD-2012, Republic of Moldova, Europe
Printed at: see last page
ISBN: 978-620-7-85970-2

Copyright © Gilmar Paulinho Triches
Copyright © 2024 Dodo Books Indian Ocean Ltd. and OmniScriptum S.R.L publishing group

Table of contents:

I dedicate this work to the Brazilian rural workers who, through their work, produce the food that sustains us, especially to my parents Adelino and Nelci, who were born farmers and have always been an example of character, humility and dignity in my life.

ACKNOWLEDGMENTS

I thank God and my parents for allowing me to be in this world walking the path I have chosen, always supporting and advising me.

With the help of many people, this work was possible.

In particular, I would like to thank Professor Dr. Anibal de Moraes, who, with his immense wisdom and typical patience, was able to build trust, respect and admiration between us, overcoming obstacles far greater than I could have imagined, for his scientific guidance, and especially for his friendship.

To Dr. Vanderley Porfírio-da-Silva for his invaluable collaboration as project mentor, encourager and motivator.

To the invaluable professors and researchers Henrique Soares Koehler, Raquel R. B. Negrelle, Cícero Deschamps, Ricardo A. de Oliveira, Flávio Zanette, Alessandro C. Angelo, Sebastiao B. C. Lustosa, for the example of educators that they are. I had the honor of being able to exchange ideas and knowledge, enriching my academic and professional training, and they will always have my admiration and gratitude.

To my postgraduate colleagues, friends and brothers: Amanda Moser C. da Fonseca, Lauri J. Marconatto, Liliane Martins, Luciano Alves, Ivete M. Guisa, Marcos A. Paladini da Silva, Romulo J. Debarba, Rodrigo M. Monzani and Teomar D. Silva.

To my friends at IFC, Joao Celio de Araújo, Veruschka R. M. Andreolla, Genuino Negri, Uberson Boaretto Rossa, among others, who inspired and supported me in many different ways.

To my parents Adelino and Nelci Triches and my brothers Gilberto, Bernardete and Rudimar, who, even though they are far away, have always supported me in my academic journey and who are the strongest link to my past.

To the Federal Institute of Education, Science and Technology of Santa Catarina, Ministry of Education, Federal Government, for the possibility of participating in the postgraduate program.

To the Postgraduate Program in Agronomy and Plant Production at the Federal University of Paraná (PGAPV/UFPR).

The Coordenapao de Aperfeipoamento de Pessoal de Nível Superior (CAPES) funded the IFC/UFPR DINTER.

To all the people who, in one way or another, contributed to the development, planning and execution of the experiment, such as Lucimara Antunes from the PGAPV coordination, Maria Emília Kudla from the phytotechnics laboratory, fellow postgraduates Leonardo Deiss, Maurício Z. Schuster, Leonardo S. Symczak, Paulo Marques, Alessandro Y. Schuster. Schuster, Leonardo S. Symczak, Paulo Marques, Aline Y. Utima, Rubia L. D. Lima, Delma F. F. Silva, Natália M. Gonzales, Édina C. P. Lopes, Breno Menezes Campus, Thales Baggio Portugal, Rafael Bonatto, Carlos Henrique Coimbra, Daniela Martin and Silvano Kruchelski.

To all the trainees at NITA, who were fundamental in contributing to the development and especially the execution of the experiment.

Finally, I would like to thank my family: my wife and partner Jacinta, my children Adriane, Giovani and Giane, my son-in-law Ailto and my grandson Rafael, a new member of the family who was born during the course of this work to brighten up the family atmosphere, who were the people who influenced me the most before and during my doctorate, and who contributed unconditionally to my psychological and emotional support.

THANK YOU ALL VERY MUCH

"Knowledge is not enough, you have to apply it.
Good will is not enough, you have to do it."

(Johann Wolfgang von Goethe, 1749-1832)

"Only knowledge that makes us better is useful"

(Socrates)

SUMMARY

Trees can improve the productivity of an agro-ecosystem, influencing soil characteristics, microclimate, hydrology and associated biological components. There is still little information on the most suitable tree species for use in integrated agricultural production systems (SIPA), especially in environmental protection areas (APA). Integrated systems have been increasingly adopted in Brazil, even supported by public policies such as the federal government's "Low Carbon Agriculture" program. These systems integrate diversification, income generation and environmental protection. However, damage to trees caused by the presence of cattle in the system can be a problem that makes it impossible to achieve technological advantages and can lead to economic losses in the use of the tree component. The causes of this damage are not precisely known and few alternatives are available to prevent or reduce it. To this end, two experiments were carried out at the Canguiri Experimental Farm - UFPR, located in the Iraí River Environmental Protection Area (APA do Iraí), in Pinhais, PR, Brazil. The first aimed to assess the effect of pruning on the debarking of the trunk of *Eucalyptus benthamii* Maiden et Cambage trees at 26 months and the damage caused by the presence of animals in a silvipastoral system. Six types and five intensities of damage were characterized. There were significant differences between the treatments with pruning (CD) and without pruning (SD) for the different types of damage. In the SD treatment, all the trees showed damage, while in the CD treatment, around 16% of the trees showed no damage. Damage of greater intensity, and therefore of greater importance for the development of the trees, was recorded in the SD treatment, where class two damage (d2=average damage) was found in approximately 85% of the trees. The practice of pruning *Eucalyptus benthamii trees* did not affect the growth in DBH and total height of the trees and contributed to reducing the incidence and intensity of damage caused by cattle. The aim of the second experiment was to assess the growth of *E. benthamii* Maiden et Cambage, from planting to 36 months of age, using a randomized block design with four treatments: cattle-forest (PF), crop-livestock-forest (LPF), crop-forest (LF) and monoculture forest or forest massif (FM). The perennial summer pasture used in the livestock systems was *Megathyrsus maximus* - cv Aries with direct sowing of black oats (*Avena strigosa* Schreb.) in winter, without desiccation. Systematic inventories were carried out with 10% sampling intensity, at 12 months, 24 months and 36 months, to measure diameter at breast height (DBH) and total height (H). There was no significant variation ($p < 0.05$) between treatments in the total height growth of *Eucalyptus benthamii* at 12 months. At 24 and 36 months, there was a significant variation ($p < 0.05$) between the treatments, with the FM treatment being superior to the other treatments. At 36 months, the average total height of the FM treatment was 14.00 m, followed by the LF (12.37 m), LPF (11.01 m) and PF (10.88 m) treatments, which had the lowest total height. When comparing average DBH growth between treatments, there was a significant difference ($p < 0.05$) only at 12 months of age, with the highest growth for the LF treatment (3.02 cm) and the lowest for the FM treatment (2.14 cm). In the evaluations at 24 and 36 months, there was no significant difference ($p < 0.05$) between the treatments.

Keywords: Eucalyptus, Integrated Systems, Herbivory, Wood.

INTRODUCTION

Integration systems (IS) basically consist of the production of crop, livestock and/or forest components produced in the same area in a systematic way, in which their allocation in space and time allows for good productivity between these components (BALBINO *et al.,* 2011). According to the same authors, IS is subdivided into 4 groups, depending on the presence of the components: Agropastoral or Crop-Livestock Integration (CLI), Silvipastoral or Livestock-Forest Integration (LFI), Silvicultural or Crop-Forest Integration (CLFI) and Agrosilvipastoral or Crop-Livestock-Forest Integration (CLFI). According to Nair (1993), production systems deliberately composed of trees (woody perennials) in association with annual crops and/or animals in some form of consortium and/or succession are called agroforestry systems.

The Brazilian territory is well suited to the spread of integration systems, as it has characteristics such as: a large number of degraded areas in agricultural crops or pastures; a high number of small properties in some regions; the presence of disorganized hydrographic basins that serve as water sources for municipalities with high population density; an increase in the rural exodus due to the intensification of agriculture based on the use of external inputs and mechanization; a significant reduction in biodiversity in areas of agricultural production; the absence of natural shading in areas of extensive livestock production; deficiencies in soil conservation practices and others.

Although the term Crop-Livestock Integration (CLI) is the most commonly found in national technical literature, Carvalho *et al.* (2014) recommend using the term Integrated Agricultural Production Systems (SIPA) for scientific publications, as it is closer to the concept adopted by the FAO and international literature. To facilitate understanding of this work, SIPA is referred to as production with crop, livestock and/or forest components produced in the same area in a systematic way.

Eucalyptus is grown for a wide variety of purposes, such as paper production, cellulose, biomass, charcoal, reconstituted wood panels, sawmills, sleepers, poles, oils for the pharmaceutical industry, cosmetics, honey, ornamentation and windbreaks (SANTOS *et al.*, 2001).

According to Jaeger and Ziger (2007), due to its good characteristics, eucalyptus is grown on a large scale in Brazil and has become a potential source of raw material, replacing wood from native forests.

By the year 2050, the world's population will increase from 7 to 9 billion people and, consequently, food production is expected to increase by around 70% (UN, 2013). On the other hand, Brazil has a large territory, which allows part of the food to be produced on Brazilian soil to meet the demand for these products. However, animal production, particularly pasture-based ruminant production, is hampered by the degraded state of Brazil's pastures. According to the latest agricultural census carried out by the Brazilian Institute of Geography and Statistics (IBGE) in 2016, of the 180 million hectares of pastures (native and cultivated), 85% are at various levels of degradation (IBGE, 2006), and as a result, they do not meet the expected productivity and generate environmental and economic damage. According to Strassburg et al. (2014), the current productivity of Brazil's cultivated pastures is 32-34% of their potential and increasing productivity to 49-52% of potential would be enough to meet the demand for meat, grain, wood products and biofuels until at least 2040, without increasing the conversion of areas with natural ecosystems.

The increase in population, *per capita* consumption and kilocalories per person will influence the increase in demand for agricultural and livestock products. The prospect is that urban growth will continue, with more people in cities depending on the products grown by fewer people in the countryside. This need to increase productivity in a sustainable way is considered to be the biggest challenge facing the agricultural sector in order to guarantee food security in 2050 (OECD/FAO, 2012).

How can we enable farmers to produce in new ways that generate income, reduce environmental impact and keep people in the countryside, the famous tripod of sustainability: economic, social and environmental?

The intensification of production in areas currently under agricultural and/or livestock use has been proposed as a solution to resolve the conflict between the expansion of areas for production and the conservation of natural ecosystems (MARTHA JÚNIOR & VILELA, 2009; VILELA et al., 2012; CORDEIRO et al., 2016). It is possible to increase the efficiency of agriculture and reduce and/or mitigate GHG emissions by conserving natural resources and improving land management (STRASSBURG et al., 2014).

SIPA is currently being proposed as a possible alternative and can be seen as a new way of planning agricultural production. It's not something new, since many producers were already working with this concept (MORAES et al., 2012), but recently these practices have become better systematized and the tree component has gained prominence, corroborating to meet environmental and productive aspects of the rural property (PORFÍRIO-DA-SILVA, 2015).

Although the number of scientific studies in Brazil seeking a better understanding of integrated agricultural and livestock production systems is growing, few are actually carried out with a systemic view, often requiring a greater temporal and spatial scale in order to detect the interactions between the various variables analyzed and the new properties that emerge from these systems (MORAES et al., 2012).

In Brazil, according to law n° 6.902/1981, the executive power, when there is a relevant public interest, may declare certain areas of the National Territory to be of interest for environmental protection, in order to ensure the well-being of human populations and to conserve or improve local ecological conditions (BRASIL, 1881).

Some activities are prohibited or restricted in these conservation units, such as industrial activities, earthmoving, mining and the use of pesticides (PARANÁ, 1996), requiring adaptations and new technologies aimed at orderly and sustainable development.

Considering the importance and need to develop more sustainable production systems, especially in APA areas, where there are still many questions about the use of the tree component combining and interacting within a system throughout the entire cycle, such as the various combinations of spatial arrangements, cultural treatments, management and soil and climatic conditions that will have an effect on nutrient cycling, diseases and pests, crop performance, as well as economic return, environmental and social contribution, this study sought to evaluate the performance of *Eucalyptus benthamii* Maiden et Cambage in monoculture SIPA, from its implantation until the third year of development and the impact of the presence of cattle in an APA area.

CHAPTER 1

LITERATURE REVIEW

The ideal tree species for integrated agricultural production systems (SIPA) should have rapid initial growth, to facilitate establishment; a reduced or not very dense crown and a long stem, which reduces shading of the intercropped crop; economic viability, offering products with high marketing potential; and low or no invasive potential, to avoid excessive propagation (DIAS-FILHO, 2006). The tree used in integrated systems must also be adapted to the region's soil and climate, as well as improving the area's microclimate (PORFÍRIO-DA-SILVA et al., 2010); it must not cause allelopathy or deposit excessive amounts of litter (MELOTTO et al., 2011); it must not be toxic to animals (PORFÍRIO-DA- SILVA & MORAES, 2010). Determining spatial planting arrangements in order to optimize the production of wood and fodder is a challenge for its establishment (MACEDO et al., 2010).

Eucalyptus (*Eucalyptus spp.*) is one of the most widely used exotic species for intercropping with crops in integrated systems, especially when intercropped with forage plants. The *Eucalyptus* genus encompasses species with high ecological plasticity, with the potential for establishment, production and adaptation to different soil and climatic conditions and, for this reason, it is the most researched forest crop and its exploitation is growing (VENTURIN et al., 2010; MACEDO et al., 2010).

In recent decades, the expansion of *eucalyptus* (*Eucalyptus spp.*) cultivation has been remarkable in Brazil, so much so that the country has risen in the international market as one of the main pulp producers (ABRAF, 2013). In Brazil, significant production gains are being achieved for the *Eucalyptus* genus through the use of appropriate silvicultural techniques, soil preparation and fertilization, combating pests and diseases and genetic improvement, with plantings of pure or hybrid varieties, as well as the cloning of materials, which maintain favourable characteristics and increase the homogeneity of stands (HIGASHI *et al.*, 2000).

Interest in silvopastoral and agrosilvopastoral systems has been increasing significantly throughout the country. Practices for integrating crops, livestock and forestry are already appearing in research and extension programs run by government agencies. These initiatives are the result of the recognition and need to conduct such systems in favor of the soil, the environment and present and future human generations (BERNARDINO & GARCIA, 2009).

EUCALYPTUS CULTURE (*Eucalyptus spp)*

The genus *Eucalyptus* belongs to the family Myrtaceae, subfamily Leptospermoideae, with around 600 species and a large number of varieties and hybrids (FOELKEL et al., 1975). With the exception of *E. urophylla* and *E. deglupta*, the other species originated in Australia (MORA & GARCIA, 2000).

The *Eucalyptus* genus originated in Australia and Indonesia and was brought to Brazil in the middle of 1825. Its use for economic purposes began in the 20th century and, to this day, the tree is used to produce cellulose, wood, synthetic fabrics and more (GALZERANO & MORGADO, 2008). According to Silva et al. (2008), eucalyptus is one of the most widely planted crops in the world, with more than 700 species, which have a high production capacity and adapt to the most diverse climate and soil environments, as well as serving different segments of industrial activity, such as: firewood, coal, cellulose, poles, furniture, among others.

Until the beginning of the 19th century, eucalyptus was planted as an ornamental and windbreak tree, and there were few industrial and forestry plants (ANDRADE, 1961).

The systematization and the enormous series of experiments are due to the Companhia Paulista de Estradas de Ferro (Paulista Railway Company) for the studies begun at the end of 1903 in Jundiaí, in the state of Sao Paulo, by the remarkable work carried out by Edmundo Navarro de Andrade, with the aim of supplying fuel for locomotives, wood for posts and sleepers (ANDRADE, 1961).

Among the many existing tree species, eucalyptus has been extensively used in forest plantations due to its fast growth, productivity, wide diversity of species, great adaptability and application for different purposes (MORA and GARCIA, 2000).

Eucalyptus also stands out on the world stage for being a fast-growing tree that can thrive in a wide range of soil and climatic conditions and can generate products for a variety of purposes, from essential oils (VITTI & BRITO, 2003) to wood for sawmills and veneers (IBÁ, 2014).

Today, the whole world recognizes the use of eucalyptus wood as an excellent fibre for producing better quality paper. As well as being economically viable, due to its rapid growth, it allows for greater productivity without harming nature (MORA and GARCIA, 2000).

The best-known species of *Eucalyptus* are trees typical of tall forests, reaching heights of between 30 and 50 meters, and of open forests, with smaller trees between 10 and 25 meters tall, including around 30 to 40 species of shrubs (MORA; GARCIA, 2000).

Eucalyptus is grown for a wide variety of purposes, including paper, cellulose, biomass, charcoal, reconstituted wood panels, sawmills, sleepers, poles, oils for the pharmaceutical industry, cosmetics, honey, ornamentation and windbreaks (SANTOS *et al.*, 2001).

The research and development activities carried out by companies in favor of genetic improvement have generated significant gains in the average annual increment (IMA). According to ABRAF (2013), the average productivity of *Eucalyptus* plantations, which was 39.4 m^3 ha^{-1} $year^{-1}$ in 2006, reached 40.7 m^3 ha^{-1} $year^{-1}$ in 2012. IBÁ (2016) cites that the average productivity of eucalyptus plantations in Brazil in 2015, as reported by forest-based companies, was 36 m3.ha^{-1} .ano^{-1} , keeping Brazil at the top of the global ranking for forest productivity. Over the last five years, eucalyptus productivity has increased at a rate of 0.7% per year.

Among the *eucalyptus* species cultivated in Brazil, *E. grandis, E. saligna and Eucalyptus urophylla* stand out, as well as plantations with hybrids, mainly E. *grandis* x *E. urophylla,* which have a large planted area mainly due to the use of clonal material (HIGA *et al.*, 2000a).

The economic benefits of agroforestry go beyond the income obtained from production, such as the possibility of financial gains from carbon sequestration (NAIR, 2012). Another important economic benefit is the fact that production is not concentrated at certain times of the year, but is better distributed throughout it

(FARREL & ALTIRERI, 2012), as well as offering diversified production without degrading the soil, making it a promising agricultural model in the alliance between agricultural production and environmental conservation (AYRES & RIBEIRO, 2010).

However, despite all the advances in *Eucalyptus* cultivation, there are certain limitations to its expansion in the southern region of Brazil, mainly due to the occurrence of lower temperatures and frequent frosts (NISGOSKI *et al.*, 1998).

The eucalyptus species that are economically important for Brazil's colder conditions are a very restricted group, *E. benthamii*, *E. dunnii*, E. *saligna*, E. *grandis,* and the hybrid between E. *benthamii* x *E. dunnii, which is* well suited to producing wood for energy and/or solid wood purposes. A second group of species should also be considered, made up of *E. viminalis*, E. *camaldulensis*, E. *cloeziana* and *E. badjensis, which are* economically important for colder conditions (PALUDZYSZYN FILHO *et al.*, 2006).

In general, the genus produces wood with a wide variety of colors (LIMA, 2005). According to Oliveira (1999) apud Rocha (2000), the wood has a low gloss, straight to coarse grain, fine to medium texture, soft to moderately hard when cut, with a distinct smell and taste, and a specific mass varying between 0.40 and 1.20 g/cm^3 .

Barcellos et al. (2005) point out that the positive aspect of eucalyptus wood is its wide range of properties, due to the diversity of species, ranging from light wood with low durability to wood suitable for a wide variety of uses.

Since eucalyptus is the fastest-growing forest species commercially planted in Brazil and, as a result, the most knowledge has been accumulated about the species, it has become the most widely used in these forestry systems. Its popularity has led many producers to believe that implementing systems with an emphasis on this species is very simple. Like any other agricultural or forestry species, eucalyptus needs to be treated with care. In order to develop it satisfactorily in monoculture, modern technologies are needed and, in order to transpose it to silvopastoral systems, it is necessary to have creative intelligence and a lot of observation on the part of producers and technicians (MEDRADO et al., 2009).

Eucalyptus as a raw material in the sawn timber industry

According to Moura and Guimaraes (2003), approximately 90 countries around the world grow eucalyptus in commercial plantations. In South American countries, the genus was introduced at the end of the 20th century, although it was only after the 1960s, due to tax incentives from the governments of Brazil and Argentina, that significant increases were seen in the area planted (JANKOWSKY et al., 2000).

The genus first spread throughout the world due to botanical and ornamental interests, and its first industrial uses were limited to energy generation and cellulose production (ACOSTA, 2006).

In Brazil, the genus occupies a prominent position in the choice of species for planting forest stands, with most of the production committed to traditional uses such as pulp and paper, charcoal, firewood and fiberboard (ROCHA, 2000).

According to the Brazilian Tree Industry (IBÁ), in 2015 eucalyptus plantations occupied 5.6 million hectares of the country's planted tree area and were mainly located in: Minas Gerais (24%), Sao Paulo (17%), Mato Grosso do Sul (15%), Bahia (11%), Rio Grande do Sul (6%) and Paraná (5%). Over the last five years, eucalyptus plantations have grown by 2.8% per year. Mato Grosso do Sul stands out with 450,000 hectares planted in the last five years (IBÁ, 2016).

According to IBÁ (2016), in 2015 the country consumed 194.45 million m^3 of wood from forest plantations, an increase of 2.2% compared to 2014, of which 151.20 million m^3 (77.76%) was eucalyptus wood, used in the different segments of the forestry industry.

According to Lobao et al. (2004), the potential for multiple uses of eucalyptus wood has been growing in recent years, and different types of forest can be obtained for each of the products to be obtained.

Jankowsky et al. (2000) state that eucalyptus is no longer an alternative raw material for the timber industry, and its transformation into solid wood products is a reality. On the other hand, they add that there is a need to adapt the wood processing processes, since the use of *Eucalyptus* implies the processing of young and small trees.

In Brazil, most eucalyptus forests are managed to produce small-diameter trees for use as fiber or energy, without much concern for the quality of the wood, and it is very difficult to use them to produce sawn timber that is transformed into furniture, window frames and other higher value-added products. At the same time, the few forests managed to supply sawn timber have not been genetically improved and often have straight trees with good diameters, but produce wood with great variation in its main properties (PONCE, 1995).

According to Amparado et al. (2008), the prejudices attributed to eucalyptus in the sawnwood industry stem from incorrect use and the lack of development of technologies to optimize the characteristics of the final product.

Despite the intrinsic problems of the genus, its wood can be used without major difficulties, as long as its limitations are taken into account, thus representing high raw material potential for the forestry industry, due to its physical-mechanical characteristics and the fact that plantations are located close to major consumer centers, reducing transportation costs (ROZAS MELLADO, 1993).

In countries such as Australia, South Africa, Chile, New Zealand, Uruguay and Argentina, Eucalyptus wood is already being used in sawn form (LIMA, 2005). However, if this type of use is to be intensified in Brazil, there is a need for more research into the subject, as is the case in these countries (MIRANDA & NAHUZ, 1999).

The proper application of eucalyptus wood, as well as other materials, is directly related to knowledge of its characteristics, which would enable greater economy and safety in the use of this material (RODRIGUES, 2002).

Eucalyptus benthamii Maiden et Cambage

Eucalyptus benthamii Maiden et Cambage is native to Australia, found in restricted areas southwest of the city of Sydney on plains along the Nepean River and its tributaries. It is a tree with a height ranging from 30 to 45 m (FAO, 2004) and a lifespan of up to 200 years. It has smooth bark, white inflorescences between March and May, its capsule is 4 to 5 mm long and the seeds are dispersed between June and December (BENSON and McDOUGALL, 1998).

In Australia it is an endangered species, occurring only in two small populations, a small population of individuals scattered along the Nepean River between Wallacia and Camden and another larger population in Kedumba Creek (33°49'S, 150°22') and upstream of the confluence with the Coxs River (BENSON, 1985). It occurs naturally on fertile soils in the flat parts of river deposits (PRYOR, 1981), in regions with an altitude of less than 100 m, with an average maximum temperature of 26 °C and an average minimum temperature of 4 °C, with the occurrence of light frosts and an average annual rainfall of 1100 mm (HIGA & PEREIRA, 2003).

In Australia, the species has shown relatively high growth rates and has been able to grow on a variety of sites, including those subject to drought and frost (JOVANOVIC & BOOTH, 2002; FAO, 2004). In Brazil, according to Higa and Pereira (2003), the results of plantations with the species are scarce and, furthermore, the uses of the wood are still being evaluated (PALUDZYSZYN FILHO et al., 2006).

However, although there are few studies on the technological characteristics of *E. benthamii* wood, there has been considerable interest in the species in southern Africa and parts of South America, where it has been used to plant forest stands for pulp and paper production (FAO, 2004).

In Brazil, the species was introduced by Embrapa Florestas in 1988, in Colombo, Paraná State, from ten matrices coming from *Wentworth Falls*, NSW (Australia) (GRAÇA et al., 1999). In the southern region, this species has shown good growth and frost resistance in experimental plantations at two and three years of age in the state of Santa Catarina (Higa, 1999). In Paraná, in the municipality of Guarapuava, according to Paludzyszyn Filho et al. (2006), *E. benthamii* was more resistant to frost than *E. dunnii, and* in the region of Dois Vizinhos, Higa and Carvalho (1990) observed 70% survival in 45-month-old plantations, an average height of 16 m and an average DBH of 15 cm, concluding that the species deserves special attention from breeders.

The planting of *E. benthamii* is recommended for regions with absolute minimum temperatures of up to -10 °C. Below this limit, there may be significant losses in its development (FAO, 1981; PALUDZYSZYN FILHO et al., 2006).

It is especially suitable in the higher regions of the states of Paraná, Santa Catarina and Rio Grande do Sul, where the average annual rainfall is between 1500 and 1900 mm with uniform distribution (HIGA *et al.*, 2012). This species has also been planted by
Klabin company, on the Santa Catarina plateau, in lower elevation regions, i.e. in the lower parts of the terrain, where cold air is concentrated and frost intensity increases (STAHL *et al.,* 2012).

Alves et al. (2011), when studying the wood of *E. benthamii* for the production of Kraft pulp, observed that the anatomical structure of the species is very similar to that of other species of the genus. In general, its wood is characterized as moderately hard; with distinct heartwood and sapwood, the heartwood being reddish brown and the sapwood pinkish yellow; with a fine to medium texture and irregular grain (NISGOSKI et al., 1998).

With regard to the species' potential uses, Higa and Pereira (2003) indicated that *E. benthamii* wood is dimensionally unstable, with high levels of volumetric shrinkage and anisotropy coefficient, which may explain the high intensity of defects in sawn timber. However, the authors point out that these studies were carried out with a small number of samples and should therefore be used with caution.

Growth, production and use of wood

In the municipality of Colombo, state of Paraná, the average height observed in *E. benthamii* from *Wentworth Falls (NSW)* at eight years of age was 21.7 m and the diameter at 1.3 m (DBH) was 18.2 cm (HIGA and PEREIRA, 2003). Also in Colombo, it grew 2.3 m in height and 2.7 cm in diameter per year, on average, at eight years of age (PEREIRA *et al.*, 2000). Also in Paraná, but in the municipality of Dois Vizinhos, the average height observed at four years of age was 16 m and DBH 15 cm (HIGA and CARVALHO, 1990).

In the north of Argentina, in the province of Jujuy, *E. benthamii* from *Coxs River* showed productivity of 34 m^3 ha-1 year-1 at seven years of age (MENDOZA, 1983).

The average individual volume in a 6-year-old *E. benthamii* stand with an initial spacing of 3 m x 2 m in the Guarapuava region, state of Paraná, was 0.527 m3 (ranging from 0.402 to 0.652 m3). In a simulation for this stand at the age of 10, the volume per hectare would be 878 m3 (LIMA *et al.*, 2011).

It is a species used for the production of firewood, charcoal, stakes, posts and mills (HIGA *et al.*, 2000a). The quality of *E. benthamii* wood indicates the species' potential use as firewood, charcoal and cellulose, but it is not recommended for sawmills, as in addition to cracks in the tops of logs and boards, it shows warping and internal cracks during drying, with high rates of volumetric contraction and anisotropy coefficient, even under mild drying conditions at room temperature and in the shade. Genetic improvement programs can partly solve these restrictive factors (HIGA and PEREIRA, 2003).

Its quality is suitable for energy production, but it presents serious problems for mechanical processing, requiring further research to reduce the intensity of cracking and warping. Its average density at seven years is comparable to *E. saligna* (PEREIRA *et al.*, 2001).

The quality of sawn timber at six years of age shows a high intensity of defects (bowing, bending, cupping and cracking), which are intensified by the drying process. Sawn timber yields are higher in the radial splitting direction (MÜLLER, 2013).

The wood is suitable for producing charcoal in conventional kilns, reaching a density of 215 $Kg.m^{-3}$, with a conversion rate of 2.4 stereos of firewood for each cubic meter of charcoal, at an age of six years (LIMA *et al.*, 2012).

Frost resistance

In Brazil, the Wentworth Falls (NSW) variety, introduced by EMBRAPA Florestas, in the municipality of Colombo, state of Paraná, showed high resistance to frost, rapid growth, good stem shape and high homogeneity of the stem. These characteristics make *E. benthamii* an excellent choice for reforestation in cold climate regions, especially in places where there are frequent and severe frosts, such as in southern Brazil (GRAÇA *et al.*, 1999).

Winter is a major problem for the establishment of eucalyptus trees in the southern region. In addition to desirable growth characteristics, the species must be cold-tolerant. In experimental plantations of two to three years, *E. benthamii* has shown good growth and resistance to frost (EMBRAPA, 1988).

The effects of temperatures close to or below zero degrees cause damage to eucalyptus trees, ranging from loss of leaf area to plant death. The younger the eucalyptus, the greater the damage to leaves, stems and branches, due to the plant's greater sensitivity as a result of its proximity to the ground, where the thermal inversion is more pronounced (PALUDZYSZYN FILHO & SANTOS, 2005).

In Paraná, the greatest limitations for eucalyptus are the low temperatures in the winter, especially the absolute minimums in some regions (PALUDZYSZYN FILHO *et al.*, 2006).

E. benthamii is a frost-resistant species that grows well (HIGA *et al.*, 2000a). The hybrid between *E. benthamii* and *E. dunnii* is an alternative in regions where frost is frequent (BRONDANI *et al.*, 2009).

The absolute minimum temperature that *E. benthamii can* withstand, recorded in a meteorological shelter, is up to -6°C. *E. benthamii* has been found to be more tolerant of frost than *E. dunnii* (DARROW, 1995).

It is suitable for planting in regions with absolute minimum temperatures of up to -10°C, under conditions of prior acclimatization by gradually lowering the temperature in the cold season. With regard to frost, it can withstand up to 25 annual frosts (FAO, 1981).

INTEGRATED SYSTEMS

Integrated agricultural and livestock production systems can guarantee the sustainable intensification of agriculture, promoting an increase in the production of food, fiber and energy, combined with the maintenance of environmental services. On the other hand, their adoption depends on the availability of information on agronomic and economic potential, market demands and government policies for sustainable development (MORAES et al., 2012).

Integration systems can be classified into four different types, according to Balbino et al. (2011): i) crop-livestock integration (CLI) or agropastoral, a production system that integrates the crop and livestock components in rotation, consortium or succession, in the same area and in the same agricultural year or for multiple years; ii) livestock-forest integration (LFI) or silvipastoral, a production system that integrates the livestock (pasture and animal) and forest components, in consortium; iii) Crop-Livestock-Forest Integration (CLFI) or silvopastoral, a production system that integrates the forestry and agricultural components through the intercropping of tree species with agricultural crops (annual or perennial); iv) Crop-Livestock-Forest Integration (CLFI) or agrosilvopastoral, a production system that integrates the agricultural, livestock and forestry components in rotation, consortium or succession, in the same area. The crop component can be restricted to the initial phase of planting the forest component or be part of the system for several years.

The term Crop-Livestock-Forest Integration (CLFI) was used by the Brazilian federal government when it developed its public policies, such as the ABC Plan (National Plan for Low Carbon Agriculture), one of the goals of which is to develop CLFI in the country (BRASIL, 2012), and with the creation of Law No. 12.805, of April 29, 2013, which instituted the National Crop-Livestock-Forest Integration Policy (BRASIL, 2013).

For SIPA in Brazil, the year 2013 became a benchmark, as Law No. 12.805, of

April 29, 2013, established the National Crop-Livestock-Forest Integration Policy (ILPF). Among other objectives, the law provides for the recovery of degraded areas and the reduction of deforestation. The ILPF is part of the Low Carbon Agriculture Programme (ABC), which is one of the sectoral plans drawn up in accordance with Article 3 of Decree 7.390/2010 and aims to organize and plan the actions to be taken to adopt sustainable production technologies, selected with the aim of responding to the commitments to reduce GHG (Greenhouse Gases) emissions in the agricultural sector made by the country to the world community.

Despite this, authors such as Carvalho et al. (2014) recommend using the term Crop-Livestock Integration (CLI) for technical literature and Integrated Agricultural Production Systems (SIPA) for scientific publications.

According to research by EMBRAPA (2016), in Brazil the area with some type of SIPA adoption covers 11.5 million hectares. The states that stand out in terms of adoption area are Mato Grosso do Sul, with two million hectares; Mato Grosso, with 1.5 million; Rio Grande do Sul, 1.4 million, which also stands out as the state with the highest number of properties participating in one of the modalities; Minas Gerais, 1.0 million, and Santa Catarina, with 680 thousand hectares. The state of Paraná only had 416,000 hectares. In addition, according to this survey, among rural producers whose main activity is livestock farming and who adopt the strategy, 83% use the crop-livestock integration system - ILP, (9% ILPF, 7% IPF or livestock-forest integration) and among grain producers, 99% adopt the crop-livestock integration system - ILP (0.4% ILPF and 0.2% ILF).

Silvipastoral systems (SSP) are associations of forest species with herbaceous or creeping forage plants and herbivorous animals that seek to sustain natural and cultivated pastures, as well as obtaining multiple plant and animal products, such as wood, meat or milk (CARVALHO et al., 1995). These are systems whose production is much more complex than that of pastures or planted forests alone. The need to maintain a balance between its components (soil, trees, fodder and animals), combined with the large number of possible interactions between these and climatic factors, increases the need for rigorous planning, which is a determining factor in successful production (BERNARDINO & GARCIA, 2009).

Agrosilvopastoral and silvopastoral systems, when established and managed properly, taking into account the economic, environmental and social aspects of rural properties, as well as the regions in which they are located, can bring benefits such as: diversifying production, increasing and better distributing income over time, conserving soil and water, providing thermal comfort for animals, improving the nutritional value of forage, ensuring greater biodiversity and sustainability for ecosystems when compared to any monocultures (BERNARDINO & GARCIA, 2009: OLIVEIRA NETO et al., 2010).

In an agroforestry system, tree species can perform different functions, such as producing wood, fruit, seeds, resin, latex and oils, as well as increasing diversity, cycling nutrients, improving the microclimate for animal husbandry and protecting the soil, among other environmental services (OLIVEIRA NETO & PAIVA, 2010). However, these authors report that wood production has often been the main justification for planting trees on rural properties. The trees to be used in integrated production systems should preferably have canopies that allow light to pass through for the growth of forage crops. It is considered that pasture production can be maintained with shading of up to 30% of the sunlight incident on the area (PORFÍRIO-DA-SILVA et al., 2010).

According to Martin (2002), the adequate availability of shade produces favourable changes in grazing behaviour and productivity, as the animals dedicate more hours a day to grazing and ruminating, food consumption is maximized under thermal comfort, the need for water is reduced and food conversion improves, with less energy used to dissipate excessive heat. The shade from trees reduces the passage of solar radiation to the animal's body surface, reducing its potential contribution to increasing the calorie load of cattle. In tropical conditions, the temperature under the tree canopy is around 2 to 3°C lower than in full sun, with reductions of up to 9°C recorded (MARTIN, 2002).

As far as forage species are concerned, in addition to being shade-tolerant, it is necessary to select species that are adapted to the management and climatic conditions of the region where they will be planted. Among the grass species that are shade-tolerant are some of the most widely used forage plants for pasture in Brazil and in other tropical and subtropical regions, such as brachiaria species and *Megathyrsus maximus* (MACEDO et al., 2010). During the establishment of the trees, the shading caused by the arboreal component is minimal and has little significant effect on the forage crop, but as the trees grow, the changes in the microclimate become more pronounced and can compromise the growth of the forage crops.

The initiative to integrate animals into forestry activities is not a recent technique. In the western United States, the practice of placing cattle to graze in native forest areas has been carried out for over 125 years (ROATH & KRUEGER, 1982). Work involving cattle in pine forests (*Pinus radiata*) in New Zealand was described by Miller (1975). In European countries, too, multiple silvopastoral farming has been carried out for many years. In Brazil, the intentional association of cattle with forests dates back to the middle of the 18th century in the region where *Araucaria angustifolia* occurs (CHANG, 1985).

In a literature review, Porfírio-da-Silva (2015) reports that research into silvopastoral systems in Paraná began in the 1980s with cattle being introduced into conventional forest plantations. A lot of research has been carried out with the aim of using the understorey as a way for wild and domestic animals (mainly cattle and sheep) to make use of it, as well as controlling the development of herbaceous vegetation, which in accumulation becomes a vehicle for spreading fires in forests (SCHREINER, 1983; BAGGIO & SCHREINER, 1988; SCHREINER, 1994; GARCIA et al., 2003) and the compatibility of the associated crop (BERNARDINO, 2007).

Crop-livestock-forest integration makes it possible to recover degraded areas by intensifying land use, allowing for greater production in a sustainable way. This system allows for high grain and fodder production, increasing the organic matter content and greater nutrient cycling, as well as diversifying rural economic activities, with the inclusion of the forestry component, such as the production of wood or energy (TRECENTI et al., 2008).

2.1.1 Use and management of eucalyptus as a tree component in SIPA

According to Porfírio-da-Silva et al. (2010), once the planning aspects for tree planting have been defined, the actual execution phase begins, when it will be necessary to prepare the area, carry out ant control, demarcate the tree planting strips, establish a pruning and tree thinning program and cultural treatments such as fertilization and control of competing plants with the tree seedlings.

Land use planning is an important procedure in the search for sustainability, with repercussions on production and the environmental aspects of rural property. In this sense, the definition of models and areas destined for production must take into account information related to soil suitability, taking into account fertility, physical structure, relief and soil and water conservation, in order to optimize productivity. In addition, the legal requirements regarding permanent preservation areas and legal reserves must be taken into account (OLIVEIRA NETO et al., 2010).

The adoption of the integration system with the tree component can be facilitated by the proper spatial distribution of the trees on the land, in order to conserve soil and water, facilitate the movement of machinery and observe the behavioral aspects of the animals. To this end, the simplest and most effective spatial arrangement is that of alleys (or renques), in which the trees are planted in strips (single or multiple rows) with wide spacing. When you want to focus on timber production, you can use smaller spacings between the strips or a greater number of rows in each strip (more trees per hectare). To focus on agricultural or livestock activities, larger spacings can be used between the alleys with a smaller number of rows (PORFÍRIO-DA-SILVA, 2006, 2007b).

When orienting the tree planting lines, first and foremost, soil and water conservation must be taken into account. Thus, the trees should be arranged on a level surface and, if terracing is necessary, the trees should be planted in the lower third of the terrace to avoid damaging the tree roots, encourage water infiltration, preserve and maintain the terrace and allow animals to move around (PORFÍRIO-DA-SILVA et al., 2010).

Wider spacing favors the development of forage in the understory and the production of larger timber, as well as allowing intercropping with agricultural crops for longer periods of time and with fewer limitations in terms of competition for space, light, water and nutrients. From the point of view of livestock farming, spacing between rows or clumps of trees can vary from 9 to 50 m, with smaller spacings limiting fodder and animal production. The spacing between trees in the row can vary from 1.5 to 5 m (DE ALMEIDA et al., 2012).

Tree arrangements can vary from single, double or triple rows, depending on the purpose of the wood, and can be associated with selective or systematic thinning practices to produce thicker wood with greater added value. In SIPA with a focus on livestock, the implementation of single rows facilitates tree management and requires less labor (DE ALMEIDA et al., 2012). Table 1 shows some examples of arrangements.

TABLE 1 - EXAMPLES OF TREE ARRANGEMENTS IN INTEGRATED SYSTEMS, ACCORDING TO INITIAL SPACING AND AFTER THINNING.

Spatial arrangement (band of trees)	Initial spacing[1]			Spacing after thinning[2]		
	Spacing (m)	Density (trees/ha)	Occupied area (%)	Spacing (m)	Density (trees/ha)	Occupied area (%)
Simple line	14 x 2	357	14,3	14 x 4 or	179 or	14.3 or
				28 x 4	89	7,1
Double line	14 x 2 x 3	417	25,0	18 x 3	185	11,1
Triple line	14 x 2 x 1.5	1.000	40,0	20 x 3	167	10,0

SOURCE: Adapted from Porfírio-da-Silva et al.(2010).
[1] Fine wood for charcoal, firewood and mills:[2] Coarse wood for sawmills and laminating.

Planting trees in SIPA when animals are present in the area can only take place if they are isolated by fences, as they can trample or trample the seedlings. Therefore, the tree component should be introduced into the system at the same time as the crop, or the area should be fenced off until the tree has developed (NICODEMO et al., 2004). The animals should enter after the first pruning, when the trees reach 6 to 8 cm in diameter at breast height (DBH) (PORFÍRIO-DA-SILVA & MORAES, 2010). At less than 6 cm in diameter, the trees are easily broken by cattle, as they are still weak and low; and this diameter meets the need to produce quality wood without knots (PORFÍRIO-DA-SILVA et al., 2010).

According to Porfírio-da-Silva and Moraes (2010), the management of the tree component, pruning and thinning, varies according to the species of tree used and its commercial purpose, as well as animal management. Thinning is usually carried out when there is intraspecific competition between the trees, detected by dendrometric monitoring.

Pruning is a silvicultural practice that consists of removing dead or living branches from trees. It is important and necessary to prune trees in order to regulate shading, maintain pasture growth and produce quality (knot-free) wood (PORFÍRIO-

DA-SILVA et al., 2010).

Thinning consists of the selective removal (cutting) of trees from the system, with the aim of: generating income, increasing the availability of light for the forage and/or agricultural component, improving the growth conditions of the trees in order to produce better quality wood and added value. Generally, thinning is aimed at maintaining a final stand of 30 to 50% of the trees planted (DE ALMEIDA et al., 2012).

Tree pruning and thinning result in changes in the quality and quantity of tree products and, likewise, in the quality and quantity of potentially available pasture. The grazing system applied and the grazing pressure control the quality and quantity consumed by the animals, as well as the amount of residue that reaches the soil. In some situations, excessive animal loads can result in greater soil compaction, which affects pasture and tree growth as a result of greater or lesser aeration and water retention (GARCIA et al., 2003).

In Cerrado regions, in ILPF systems, eucalyptus can be thinned successively, according to the purpose of the wood: at 4-5 years, for charcoal, firewood and mills and at 8-9 years, for pulp and poles, with the final cut being made at 12-14 years, for sawmills and laminations, followed by a new cycle of tree planting. It is also possible to opt for just one thinning, leaving between 100 and 200 trees/ha for the final cut, with the wood being sold to sawmills (DE ALMEIDA et al., 2012).

When harvesting trees, technical procedures must be followed for felling the tree, dragging the log, trapping the log, loading the log(s) and transporting the wood, according to the market that buys the wood and the available labor and machinery (PORFÍRIO-DA-SILVA et al., 2010).

Advantages of using eucalyptus in SIPA

Pulrolnik et al. (2010), evaluating the adaptation and initial development of eight native tree species and two exotic tree species in crop-livestock-forest integration systems, observed that the eucalyptus species showed greater growth in height and diameter compared to the native species. The same authors also compared the initial development of *Eucalyptus urograndis,* with spacing of 12 and 22 m between the crop rows, and found no difference between the different arrangements.

Eucalyptus is a plant that throughout its natural evolution has developed efficient adaptive mechanisms to grow quickly under favorable conditions and also withstand water, temperature and nutritional stresses, among others, which explains the large number of species in nature and their dispersion in their regions of origin. In Brazil, over the last 50 years, as a result of genetic improvement work and the use of advanced silvicultural techniques, biomass production capacity has grown significantly. The versatility of the use of wood, the main product obtained from this tree, is another feature that explains the prevalence of eucalyptus plantations over other species. Uses range from purely for energy purposes to the production of fine products such as veneers and furniture (PALUDYSZYN FILHO et al., 2006).

One of the main resources used by cattle managed at pasture in an attempt to mitigate high temperatures and high radiation is the search for shade, when it exists (GLASER, 2008). The adequate availability of shade produces favourable changes in grazing behaviour and productivity: animals dedicate more hours a day to grazing and ruminating; food consumption is maximized under thermal comfort; the need for water decreases; food conversion improves, with less energy used to dissipate excessive heat (MARTIN, 2002). The tree component can influence animal welfare and performance. Eucalyptus is one of the most widely used species in silvopastoral systems in Brazil, due to its fast-growing characteristics and multiple-use timber (VENTURIN et al., 2010).

According to Pofírio-da-Silva et al. (2010), the results already obtained by research and rural extension institutions, as well as in rural producers' areas, point to the following prospects for the use of silvopastoral systems: i) an increase in the stocking rate of pastures compared to conventional livestock systems; ii) promotion of erosion control; iii) less frequent pasture renovations; iv) better conditions for animal performance (reproduction and production) by increasing the comfort and protection of animals and pastures: v) elimination of investment in the construction of artificial shade for animals, especially for dairy cattle; vi) greater supply of fodder with higher nutritional value as a result of improved soil fertility; vii) production of wood for sawmills and veneer mills.

In addition to plant and animal production, optimizing production factors in time and space, the production of integrated systems with trees can make major contributions in terms of recovering degraded areas, maintaining and restoring forest cover, promoting and generating employment and income, adoption of good agricultural practices, improvement of social conditions, adaptation of rural property to environmental legislation and valuation of environmental services within the rural property and even on a larger scale with carbon sequestration and mitigation of greenhouse gas emissions (PULROLNIK et al., 2010).

SIPAS are viable in many regions of Brazil. However, maximum productivity of the components of these systems should not be expected, but rather the sustainability of the ecosystem, and the generation of satisfactory returns and income from different products (BERNARDI, et al., 2014).

Limitations on the use of eucalyptus in SIPA

In addition to the direct and indirect benefits of silvopastoral systems, it is important to note that this system is not a panacea for the various problems inherent in tropical pastures. For example, the presence of trees and shrubs in the pasture can also hinder pasture development. This is mainly due to the shading and, in some cases, competition for water and nutrients that the tree and shrub species exert on the herbaceous forage plants in the pasture (DIAS FILHO, 2006).

Silvopastoral systems are constantly touted as the solution to various problems inherent to pastures in Brazil, but in practice, the use of this system in the country is still very limited. There are many reasons for the discrepancy between the technical evidence and the practical reality. An important barrier to the adoption of silvopastoral systems would be the limited initial profitability due to the higher implementation costs and the consequent low rate of return of these systems, when compared to traditional pasture systems (DIAS FILHO, 2006). In addition, according to Dias Filho (2006), the implementation and maintenance of the tree system requires more skilled labor and a more elaborate infrastructure (such as the production of seedlings and tree pruning). These factors could also restrict its adoption. Currently (2017), some incentive policies that did not exist in 2006 have been implemented, such as the development of the

National Plan for Low Carbon Agriculture (ABC Plan) (BRASIL, 2012) and the establishment of the National Crop-Livestock-Forest Integration Policy - Law No. 12.805 of 2013 (BRASIL, 2013). These and other measures have made it possible to increase the areas with SSP in Brazil (EMBRAPA, 2016), but they are still very limited compared to the existing possibilities.

SIPA perspectives and the use of eucalyptus

The integrated system with trees is dynamic and complex, due to the interactions between crops, animals and various practices. Because they are dynamic, these systems require continuous scientific and technological research, almost always carried out through long-term, regionalized experiments, without which their sustainability would be compromised, making it difficult for farmers to adopt them. Many lines of research, technology transfer and development need to be strengthened in order to increase the supply of knowledge and information and the adoption of SIPA as a sustainable proposal (BALBINO et al., 2011).

Currently, there are incentives and specific credit lines for SIPA in Brazil, but there is still a lack of information and qualified professionals to work with the tree component integrated into production systems. Crop-livestock-forest integration systems are technically, environmentally and socio-economically viable alternatives for recovering and intensifying the use of pastures, but they are more complex, require the interaction of several areas of knowledge and higher initial investments (DE ALMEIDA et al., 2012).

It is of fundamental importance to expand scientific studies on spatial arrangements and their effects on the forest productivity of different tree species in consortium with different annual crops. In addition, studies are needed on forage species and consortium systems of annual crops, forage and trees, to analyze their influence on biomass production in the presence of animals. Socio-economic and environmental evaluations of different ways of integrating the tree component in Brazilian biomes are therefore necessary (BALBINO et al., 2011).

The tree component in integrated production systems has a major influence on animal welfare and performance while grazing. Eucalyptus, in turn, has great potential for use in such systems, allowing for increased productivity and income generation. However, more studies are still needed on the interaction between animal-plant-soil components over time (DA SILVA & DE LIMA MEIRELES, 2013).

Although there are examples of the use of SIPA in Brazil, the diversity of regional conditions in the country indicates the need for regionalized studies on the feasibility of combining different species. It is necessary to create public policy mechanisms so that farmers can overcome economic barriers, such as the need for initial investment, and operational barriers, such as the need for technological knowledge and investment in training technicians and professionals in higher education and vocational schools in the agricultural area. Therefore, multidisciplinary studies and the introduction of subjects with a cross-cutting approach to ISAP in agricultural courses should seek to understand the soil-plant-animal-environment relationship in order to establish sustainable management in the various modalities of the system (BALBINO et al., 2011).

HISTORY OF THE AREA AND EXPERIMENTAL DESCRIPTION

Location of the experimental area

The research that resulted in this work was carried out at in the municipality of Pinhais-PR, in the experiment developed by the Agricultural Technology Innovation Center (NITA), located at the Canguiri Experimental Farm, belonging to the Federal University of Paraná (Figure 1). The approximate central geographic coordinates are: 25°23'30"S latitude and 49°07'30"W longitude, with an average altitude of 920m. The total area of the SIPA experiment developed by NITA is 33 hectares, of which approximately 50% (18 hectares) is made up of the *Eucalyptus benthamii* Maiden et Cambage tree component.

The Canguiri Experimental Farm covers 444.78 hectares and is located within the Iraí River Environmental Protection Area (APA), on the banks of the Iraí reservoir. The Iraí APA is a territorial unit created in 1996 to "protect and conserve the environmental quality of the natural systems existing there, especially the quality and quantity of water for public supply purposes", as stated in Article 2 of State Decree No. 1.753/96. Conventional allotments and the use of pesticides are not allowed in these areas.

FIGURE 1 - SPATIAL LOCATION OF THE STUDY AREA

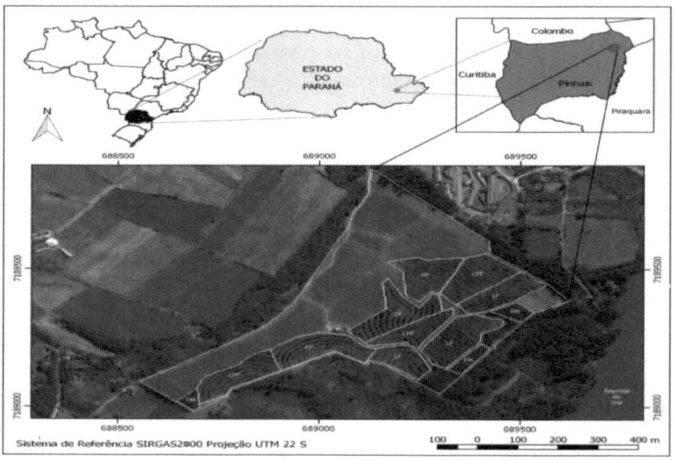

SOURCE: GOOGLE EARTH (2016), adapted by the AUTHOR (2017)

Characterization of the area

The region's climate, according to the Koppen classification, is Cfb (temperate, humid with temperate summers), characterized by an average annual rainfall of 1400 mm, an average minimum temperature of 12.5°C and an average maximum temperature of 22.5°C, and is subject to severe frosts (more than five per year).

The average data and curves for rainfall and average temperature indicators for the period 2013-2016 in the experiment area are shown in Figure 2.

According to soil mapping, the classes Cambissolo Háplico, Latossolo Vermelho and Latossolo Vermelho- Amarelo are found in this area (SUGAMOSTO, 2002).

FIGURE 2 - RAINFALL AND AVERAGE TEMPERATURE IN THE AREA

EXPERIMENTAL, IN THE PERIOD 2013-2016

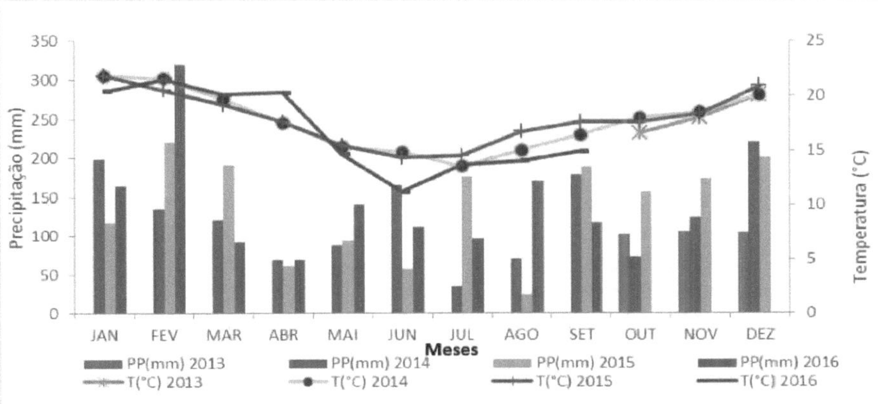

SOURCE: SIMEPAR (2016), adapted by the AUTHOR (2017).

Experiment details

The experiment began in 2012 with conventional soil preparation with deep plowing with a moldboard plow and subsequent harrowing (June and July); top dressing with sewage sludge treated by the N VIRO process® (July/August); sowing black oats (*Avena strigosa* cv IAPAR 61) at a seeding rate of 80 $kg.ha^{-1}$ to cover the soil and applying 450 $kg.ha^{-1}$ of simple superphosphate in the row.

In January and February 2013, the soil was conventionally prepared for sowing perennial summer pasture (*Megathyrsus maximus* - cv Aries). *Megathyrsus maximus* was sown in rows using 16 $kg.ha^{-1}$ of pelletized seeds and 350 $kg.ha^{-1}$ of simple superphosphate was applied in the row, plus 800 $kg.ha^{-1}$ of ammonium chloride. During the establishment of the pasture, three mowings were carried out to reduce the competitive pressure of the weeds.

The seedling *Eucalyptus benthamii* was planted in October 2013, and all the treatments received the same soil preparation and tree fertilization. The planting line was subsoiled to a depth of approximately 40 cm, and the seedlings were planted by hand using a weeder to open the holes. For each seedling, 250 ml of Hidroplan-EB®/HyB hydrogel solution was placed in the hole, diluted according to the manufacturer's recommendations, and 200 g of the 08-20-20 NPK formulation, which also contained 1% calcium (Ca) and 3% sulphur (S), was distributed in two lateral pits (15 cm away from the seedling and 15 cm deep).

In the PF, LPF and LF treatments, the trees were planted in single rows with a spacing of 2 m x 14 m, with the aim of obtaining an initial density of approximately 357 trees ha^{-1} and a final density of 150 trees ha^{-1} . The rows of trees were planted on contour lines in order to conserve soil and water, control erosion, improve the thermal comfort of the animals and soften the trails made by animals walking along the slope of the land (PORFÍRO-DA- SILVA et al., 2010). The monoculture tree treatment (FM) was planted at a spacing of 2 m x 3 m (1,667 trees.ha).$^{-1}$

Ants were controlled by applying Formifu® paste to all the seedlings fifteen days after planting (October 2015).

In the process of planting the trees, it was necessary to replant approximately 20.5% of the trees in the experiment and, as the use of herbicides is not permitted, weed control was carried out manually with a hoe, grubbing and three times with a motorized brush cutter.

The fertilization of the production system was standardized and based on soil analysis and the demands of the most demanding crop, in this case maize (*Zea mays*).

In the first year (2013), the fertilizers were applied in three stages in all the treatments, the first as a base, when the annual summer crops (corn and sunflower) were planted, using 450 $kg.ha^{-1}$ of simple superphosphate, the second with 400 $kg.ha^{-1}$ of the NPK 25:00:25 formula when the plants had 4 to 6 leaves, and a fortnight later 200 $kg.ha^{-1}$ of urea was applied.

[a]In the second year (2014), the fertilizers were applied to the treatments in three stages: the first after planting the winter pasture (oats), in the 2nd half of May, using 600 $kg.ha^{-1}$ of natural phosphate, 200 $kg.ha^{-1}$ potassium chloride and 200 $kg.ha^{-1}$ of urea; the second at the time of planting the annual summer crops (corn and sunflower) in September, using 450 $kg.ha^{-1}$ of simple superphosphate and the third in November, applying 450 $kg.ha^{-1}$ of urea.

In 2015, the fertilizers were applied in two stages: the first, in the 2nd half of May after planting the winter pasture (oats), using 300 $kg.ha^{-1}$ of natural phosphate, 100 $kg.ha^{-1}$ potassium chloride and 300 $kg.ha^{-1}$ of urea; the second applying 200 $kg.ha^{-1}$ of urea at the end of December.

In 2016 (the fourth year), after planting the winter pasture (oats), 300 $kg.ha^{-1}$ of natural phosphate, 100 $kg.ha^{-1}$ potassium chloride and 300 $kg.ha^{-1}$ of urea were applied to all treatments in the second half of May.

Four treatments were tested: T1- livestock-forest system (PF); T2- crop-livestock-forest system (LPF); T3- crop-forest system (LF), with half the area sown with corn and the other half with sunflower, and T4- monoculture forest (FM). The experimental design was randomized blocks with 3 replications (paddocks), consisting of plots with areas of at least 1.3 hectares and a maximum of 2.2 hectares in the PF, LPF and LF treatments and 0.3 to 0.5 hectares in the FM treatment (Figure 3).

In the crop-forest treatment (LF), corn *(Zea mays)* and sunflower *(Helianthus annuus)* were grown in the spring-summer periods, sown in a no-till system at the end of September and beginning of October, without the use of herbicides or any other type of agrochemical due to the fact that the experiment is located in an Environmental Protection Area (APA). In the autumn-winter periods, black oats (*Avena strigosa* Schreb.) were sown in a no-till system at the end of May and beginning of June, with the aim of keeping the soil covered with vegetation during the period.

In the PF and LPF treatments, with the perennial summer pasture (*Megathyrsus maximus* - cv Aries), in the autumn-winter periods, black oats (*Avena strigosa* Schreb.) were over-seeded in late May and early June, in a no-till system, to provide pasture in winter or to keep the soil covered with vegetation during the period in the LPF treatment, when a summer crop was grown.

FIGURE 3 - DISTRIBUTION OF *E. benthamii* IN THE EXPERIMENTAL AREA OF THE FARM CANGUIRI - UFPR, PINHAIS-PR, 2017.

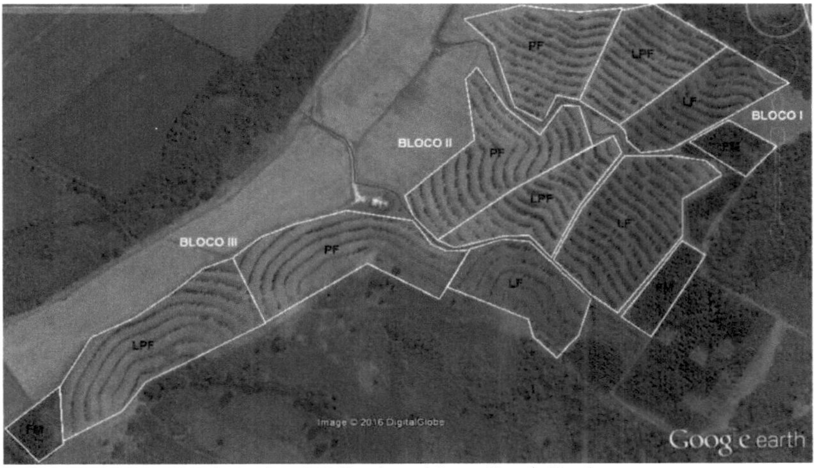

SOURCE: the author (2017), adapted from *Google Earth*.

The trees were first pruned at 21 months of age (July 2015) to half their respective heights (on average up to 3 m), using pruning saws. [a]The 2nd pruning was carried out at 33 months of age (July 2016) to half of the trees' respective trunk heights (5 to 6 m). In the FM treatment, natural pruning took place during the 2nd pruning with the death of the lateral branches, and it was only necessary to prune the trees at the edge of the plots.

The animals entered the experiment in the PF and LPF treatments at 24 months of tree age (October 2015), when the trees had an average DBH of 8.12 cm, considered adequate to support their presence (MEDRADO et al. 2009; PORFIRIO- DA-SILVA et al. 2012).

The animals used for spring-summer grazing were common (no defined breed), with an average live weight of 285 kg (± 54 kg) and ages ranging from 01 to 03 years. They were distributed in the paddocks based on their average weight, in order to keep the animal load proportional to the area of the paddocks. Grazing began in October 2015 and ended in April 2016. It was managed under continuous grazing with a variable stocking rate (MOTT & LUCAS, 1952), keeping the pasture at an average height of 28 cm (±5 cm). The cattle had access to the paddocks with *Megathyrsus maximus* - cv Aries and *Eucalyptus benthamii*, with the other factors of the grazing period (climate, soil, management, diet) constantly maintained.

CHAPTER 2
damage caused by cattle to eucalyptus benthamii trees IN SYSTEMS SILVIPASTORIS

SUMMARY

Silvopastoral systems are being adopted in Brazil, even supported by public policies such as the federal government's "Low Carbon Agriculture" program (MAPA/ACS, 2012). These systems integrate diversification, income generation and environmental protection. However, damage to trees caused by the presence of cattle can be a problem that makes it impossible to achieve technological advantages, as well as economic losses from the use of the tree component. The causes of this damage are not precisely known and few alternatives are available to prevent or reduce it. The aim of this study was to evaluate the effect of pruning on the debarking of the trunks of *Eucalyptus benthamii* trees at 26 months of age and the damage caused by the presence of animals in a silvopastoral system in an experiment at the Canguiri Experimental Farm - UFPR, located in the Iraí River Environmental Protection Area (APA do Iraí), in Pinhais, PR, Brazil. Six types and five intensities of damage were characterized. There were significant differences between the treatments with pruning (CD) and without pruning (SD) for the different types of damage. In the SD treatment all the trees showed damage, while in the CD treatment around 16% of the trees showed no damage. Damage of greater intensity, and therefore of greater importance for the development of the trees, was recorded in the SD treatment, where class two damage (d2=average damage) was found in approximately 85% of the trees. The silvicultural practice of pruning *Eucalyptus benthamii trees* did not affect the growth of DBH and total height of the trees, proving to be efficient in producing quality wood (without knots).

Key words: Forest Livestock System, Herbivory, Pruning, Wood

INTRODUCTION

Silvopastoral systems (SSP) represent a land use technology that can increase biodiversity and sustainability in agroecosystems when compared to any monoculture (BERNARDINO & GARCIA, 2009), as the integration of livestock (pastures and cattle) and forestry components is an alternative for incorporating wood production into livestock enterprises, bringing together the economic advantages that each system has separately, i.e. the quick return from livestock activities and the favorable characteristics of the market for timber forest products (PORFÍRIO DA SILVA et al., 2010).

Interest in silvopastoral and agrosilvopastoral systems has been increasing significantly throughout the country. Practices for integrating livestock and forestry are already appearing in research and extension programs run by government bodies, such as the Brazilian federal government's "Low Carbon Agriculture - ABC" program, created in 2010, and the "Integra SP" project in the state of Sao Paulo, created in 2013.

The silvopastoral system is a strategy that integrates livestock production and timber production in the same area. The association of trees, properly managed, with pastures can provide an extra source of income for livestock farmers, improve environmental protection, improve conditions for animal welfare and increase soil fertility and pasture production (CARVALHO et al., 2002; DIAS et al., 2007; BERNARDINO & GARCIA, 2009; PORFIRIO-DA-SILVA et al., 2010).

SSP have proven to be viable for meeting the demand for wood-based products, bringing in additional revenue, however, maximum productivity of the components of these systems should not be expected, but the sustainability of the ecosystem as a whole (BERNARDI et al., 2014).

The establishment of tree seedlings in the middle of the pasture is a critical phase, requiring specific care since any damage caused during this phase of the trees' development can jeopardize the success of the system (PORFIRIO-DA-SILVA et al. 2009; GARRET et al., 2004; LEHMKUHLER et al. 2003). However, even after the trees have grown enough to withstand the presence of cattle, damage to the bark can still be caused by cattle (MEDRADO et al., 2009; PORFIRIO-DA-SILVA et al., 2012; GUERREIRO et al., 2015). Damage to the bark of tree trunks has the effect of depreciating the quality of the wood and can even lead to the death of trees. The factors or causes for the removal of eucalyptus bark by cattle are largely unknown. Among the hypotheses suggested to explain this type of behavior would be nutritional imbalance, ranging from minerals, protein and energy, to the concentration and quality of fibers ingested (MALAFAIA et al., 2011; GILL, 1992); aspects such as the presence of endoparasites (SAINT-ANDRIEUX et al., 2009; VILLALBA et al., 2013; VILLALBA et al., 2014); and also behavioral aspects originating from learning (WEMELSFELDER, 1984; BERRIDGE & KRINGELBACH 2008) and/or stress (MASON et al., 2007; ANDERSON et al., 1985).

According to Ralphs & Provenza (1999), when animals in a herd observe another animal eating a certain food, this can induce other animals to do the same; then, if there is no adverse effect, all the animals will continue to eat the new food. When ingesting a plant or parts of a plant, the animal develops an ingestion or aversion reaction, which allows herbivores to differentiate between foods in order to optimize their intake of nutrients and toxins (WESTOBY, 1974; PROVENZA et al., 2003).

Nolte (2003) also considers that the number and density of animals, the availability of alternative food sources other than pasture and trees and the palatability of trees in relation to the alternative food source could all be factors that lead cattle to eat leaves, branches and/or bark.

On the other hand, Medrado et al. (2009) suggest that the wrong management and poor nutrition of trees can cause imbalances in the chemical composition of the bark which, in addition to diseases that affect the bark itself, could be the cause that leads cattle to ingest bark and cause damage to the tree trunk. Injuries to the trunk, including those caused by disease, cause the exudation and deposition of photoassimilates derived from the parenchymatous tissues to heal the injury (ROMERO, 2006), which can increase its attractiveness to animals (PORFÍRIO- DA-SILVA, et al., 2012).

Pruning is a silvicultural practice recommended for the production of logs and to mitigate the competition for light caused by tree canopies. However, pruning causes wounds that exude photoassimilates and could make the bark of the tree trunk more attractive; on the other hand, Gill (1992) considers that the absence of pruning, by keeping the branches and twigs lower, could make it more difficult for animals to access the bark of the trunk.

This statement encouraged the hypothesis that pruning could increase the risk of debarking by cattle. On the other hand, pruning has been suggested to reduce branching, as well as to prevent the breaking of branches and twigs (PORFIRIO-DA-SILVA et al., 2012) in silvopastoral systems, since the breaking of branches can cause wounds that allow pathogens to enter and the formation of 'loose knots' that depreciate the quality of the wood.

The aim of this study was to evaluate the effect of pruning on debarking caused by cattle in a silvopastoral system with *Eucalyptus benthamii* at 26 months of age.

MATERIALS AND METHODS

The experiment is part of a project developed by the Agricultural Technology Innovation Center (NITA), located at the Canguiri Experimental Farm (25°23'30"S; 49°07'30" W) belonging to the University of São Paulo.

Federal University of Paraná, located in the municipality of Pinhais-PR, with an average altitude of 920m. The farm is located within the Irai River Environmental Protection Area (APA) (State Decree No. 1.735/96), on the banks of the Iraí reservoir.

According to Koppen, the region's climate is classified as Cfb (temperate, humid with temperate summers), characterized by an average annual rainfall of 1400 mm, an average minimum temperature of 12.5°C and an average maximum temperature of 22.5°C, and is subject to severe frosts (IAPAR, 2000).

According to soil mapping, the classes Cambissolo Háplico, Latossolo Vermelho and Latossolo Vermelho- Amarelo are found in this area (SUGAMOSTO, 2002).

The NITA experiment has four treatments: T1- livestock-forest system (PF); T2- crop-livestock-forest system (LPF); T3- crop-forest system (LF), with half the area sown with corn and the other half with sunflower, and T4- monoculture forest (FM). The experimental design was randomized blocks with three replications (paddocks), consisting of plots with areas of at least 1.3 hectares and a maximum of 2.2 hectares in the PF, LPF and LF treatments and 0.3 to 0.5 hectares in the FM treatment. For this study, we used areas from treatments with a livestock component.

The experiment began in 2012 with conventional soil preparation work involving deep plowing with a moldboard plow and subsequent harrowing (June and July); soil correction with sewage sludge treated by the N VIRO process® (July/August); sowing black oats (*Avena strigosa* cv IAPAR 61) at a seeding rate of 80 kg.ha^{-1} to cover the soil and applying 450 kg.ha^{-1} of simple superphosphate in the row.

In January and February 2013, the soil was conventionally prepared for sowing perennial summer pasture (*Megathyrsus maximus* - cv Aries). *Megathyrsus maximus* was sown in rows (16 kg.ha^{-1}) and 350 kg.ha^{-1} of simple superphosphate was applied in the row, plus 800 kg.ha^{-1} of ammonium chloride. During the establishment of the pasture, three mowings were carried out to reduce the competitive pressure of the weeds.

In September 2013, the tree lines were demarcated and the seedlings of *Eucalyptus benthamii were planted* in October 2013. This species was chosen mainly because it is frost-resistant and the aim is to obtain logs for sawmills and rolling mills at the end of the 20-year cycle.

The eucalyptus trees were planted in single rows with a spacing between them of 14 m between rows and 2 m between plants, in order to obtain an initial density of approximately 357 trees per hectare, planted on contour lines to conserve soil and water, control erosion, improve the thermal comfort of the animals and soften the trails made by animals walking along the slope of the land (PORFIRIO-DA-SILVA et al., 2010).

Planting was carried out by hand using a weeder after subsoiling to a depth of 40 cm in the planting line. Each seedling received 250 ml of Hidroplan-EB®/HyB hydrogel solution, as recommended by the manufacturer, and in two lateral pits (15 cm away from the seedling and 15 cm deep) 200 g of NPK (08-20-20) containing 1% calcium (Ca) and 3% sulphur (S).

Fertilization of the production system was based on soil analysis.

In the first year (2013), the fertilizers were applied in two stages: the first in September, using 450 $kg.ha^{-1}$ of simple superphosphate and 400 $kg.ha^{-1}$ of the NPK 25:0:25 formulation, the second in October, applying 200 $kg.ha^{-1}$ of urea as a top dressing.

[a]In 2014, fertilizers were applied in three stages: the first, in the 2nd half of May, after planting the winter pastures (oats), using 600 $kg.ha^{-1}$ of natural phosphate, 200 $kg.ha^{-1}$ potassium chloride and 200 $kg.ha^{-1}$ of urea; the second in September using 450 $kg.ha^{-1}$ of simple superphosphate and the third in November, with 450 $kg.ha^{-1}$ of urea applied as a top dressing.

In 2015, the fertilizers were applied in two stages: the first, in the second half of May after planting the winter pastures (oats), using 300 $kg.ha^{-1}$ of natural phosphate, 100 $kg.ha^{-1}$ potassium chloride and 300 $kg.ha^{-1}$ of urea; the second, applying 200 $kg.ha^{-1}$ of urea at the end of December as a top dressing.

Being in an APA area, where the use of herbicides is not permitted, weed control was carried out manually by means of uprooting, the use of hoes and a motorized brush cutter during the first 120 days after planting the seedlings. In the process of planting the trees, it was necessary to replant approximately 20% of the trees in the experiment.

Leaf-cutting ants were controlled by applying Formifu® paste to all the seedlings fifteen days after planting. Orange peels were also used and distributed weekly at strategic points in the area and on its outskirts to attract the ants and locate their nests, which were then destroyed mechanically.

In July 2015, when the average DBH of the trees had already exceeded 6 cm in diameter, considered to be the minimum size needed to withstand the presence of cattle without breaking the tree trunk (MEDRADO et al., 2009; PORFIRIO-DA-SILVA et al., 2012), the first pruning of the trees was carried out by removing branches up to half the respective heights of each tree.

The animals entered the experiment when the trees were 26 months old (December 2015), when they had an average DBH of 10.41 cm.

The experiment area of 4.8 hectares was divided into three paddocks, each covering between 1.3 and 1.8 hectares (Figure 4), using electric fencing. In all the paddocks, the animals were provided with free water and mineralized salt.

The animals used had no defined breed, an average live weight of 285 kg (± 54 kg) and ages ranging from 01 to 03 years. They were distributed in the paddocks based on their average weight, in order to keep the animal load proportional to the area of the paddocks.

Grazing began on 27/11/2015 and ended on 13/04/2016, for a total of 138 days. It was managed under continuous stocking grazing with variable animal load (MOTT & LUCAS, 1952), keeping the pasture at an average height of 28 cm (± 5 cm).

A 10% systematic inventory was carried out on the trees before the animals entered and when they left, measuring the diameter at breast height (DBH) and total height at the start and end of grazing. A plot with 20 trees was also demarcated in each paddock, on which dendrometric belts were installed to monitor the monthly increment.

FIGURE 4 - DISTRIBUTION OF *E. benthamii* IN THE EXPERIMENTAL AREA OF CANGUIRI FARM - UFPR, PINHAIS-PR, 2015.

SOURCE: the author (2017), adapted from *Google Earth*.

The average DBH and average height of these trees in December 2015, at 26 months of age, were respectively: 10.23 cm and 9.37 m in the treatment without pruning (SD) and 10.59 cm and 9.15 m in the treatment with pruning (CD).

The damage caused by the presence of cattle on the trees was assessed at the end of the grazing period (April 2016). The methodology described by Porfirio-da-Silva et al. (2012) was used to assess the damage. The damage caused by the animals was categorized according to the damaged parts (incidence) of the affected plant: Tq = breakage of the main stem or trunk; Gq = breakage of secondary branches; Rq = breakage of thin branches and foraging of leaves, or ramoneio; Tl = damage to the trunk reaching the wood by removing cambial tissue; Cl = damage to the bark, without reaching the cambium; and, Dl = damage greater than 5 cm in diameter.

A system of weights and scores was established for the different types of damage: Tq = 10.0; Tl = 4.0; Gq = 2.0; Cl = 1.5; Rq = 1.0; and, Dl = 1.0. The criterion adopted according to Porfírio-da-Silva et al. (2012) seeks to relate the type of damage and its importance for the future development of the tree. The final score for each individual is the sum of the respective damage incidents, except for damage Tq (trunk breakage), whose incidence, regardless of the occurrence of other types of damage, has a maximum score of 10. The occurrence of Dl (damage to the bark greater than 5 cm in diameter), Tl (damage to the trunk reaching the wood through the removal of cambial tissue) and Gq (breakage of secondary branches) is always associated with other types of damage.

According to the criteria adopted, five damage intensity classes (d) were established: d0 = nil (no damage, score 0); d1 = low (scores between 0 and 3); d2 = medium (scores between 3 and 6); d3 = high (scores between 6 and 10); and, d4 = extreme (score 10). Data was collected from all the trees in the experiment (1,142 plants) in order to quantitatively describe and qualify the damage suffered by the different treatments.

The experimental design was randomized blocks with two treatments: T1- Livestock-Forest Integration with pruned eucalyptus (CD) and T2- Livestock-Forest Integration with unpruned eucalyptus (SD) and three replications (paddocks) of 1.3 to 1.8 ha each. The six types of damage (Tq, Tl, Gq; Cl; Rq, Dl) caused by the animals were compared between and within each treatment (CD and SD) using the Tukey test at $p<0.05$.

The mean damage scores (d0, d1, d2, d3 and d4) between and within each treatment (CD and SD) were also compared using the Tukey test at $p<0.05$.

The homogeneity of the variance of the sum of the damage was analyzed using Bartlett's test to verify the normality of the data as a prerequisite for using analysis of variance to evaluate the damage within and between treatments. ASSISTAT version 7.7 beta was used for the statistical analysis.

3.2 RESULTS AND DISCUSSION

There were significant differences for the different types of damage between the treatments with pruning (CD) and without pruning (SD) on *E. benthamii*. In the treatment with pruning (CD), the damage with the highest incidence was bark damage, without reaching the cambium (Cl) with 80.5%. In the treatment without pruning (SD), the damage with the highest incidence was secondary branch breakage (Gq) and leaf foraging (Rq), with incidence in 100% of the trees. The SD treatment also had a high incidence of bark damage without reaching the cambium (Cl) with 89.4% and bark damage greater than 5 cm in diameter (Dl) with 72.3%.

The susceptibility of eucalyptus to damage by cattle was previously reported by Medrado et al. (2009), who indicated widespread damage to *E. grandis* at three years of age in a silvopastoral system with Uruchloa brizantha cv. Xaraés in the state of Paraná, Brazil. Damage in 61% of the trees increased to 92% within a year and reached the wood in 30% of the trees.

Guerreiro et al. (2015), in an assessment of the susceptibility of 10 eucalyptus species or clones to damage caused by cattle in a silvopastoral system in the municipality of Brotas-SP, observed that there were significant differences in the level of damage between the various eucalyptus species or clones, with the most affected being *E. grandis,* with 58% of the wounds caused to the bark reaching the wood.

Porfírio-da-Silva et al. (2012) assessed three tree species for damage caused by grazing cattle in a silvopastoral system in subtropical Brazil. The species *Schinus terebinthifolius* (Raddi), *Grevillea robusta* (*A. Cunn.* ex R.Br) and *Eucalyptus dunnii* (Maiden) were planted in single rows, arranged at 14 m x 3 m, to form an agroforestry system. When the trees were 41 months old, cattle were introduced to the area and the damage caused to the bark of the tree trunks in the system was greater than that caused to the crowns. The *S. terebinthifolius* trees were the most damaged by the cattle, leading the authors to conclude that the species should not be recommended for a silvopastoral system.

In this study, damage was concentrated on leaves, branches and bark. Trunk breakage (Tq) was only observed in trees with a DBH of less than 6 cm, with an incidence of 1.8% in the treatment with pruning (CD) and 3.0% in the treatment without pruning (SD). This coincides with the result obtained by Porfirio-da-Silva et al. (2012) who only reported trees with a DBH of less than 6 cm being broken by cattle. There was an almost 100% incidence of ramoneio (Rq) and branch breakage (Gq) in the treatment without pruning (SD), while in the treatment with pruning (CD), the incidence of this damage was 7.6 and 5.7% (Figure 5).

FIGURE 5 - INCIDENCE AND TYPES OF DAMAGE CAUSED BY BOVINE IN *E. benthamii* IN THE SILVIPASTORIL SYSTEM WITH DISRUPTION (CD) AND WITHOUT DISRUPTION (SD), FAZENDA DA UFPR, PINHAIS-PR, 2016.

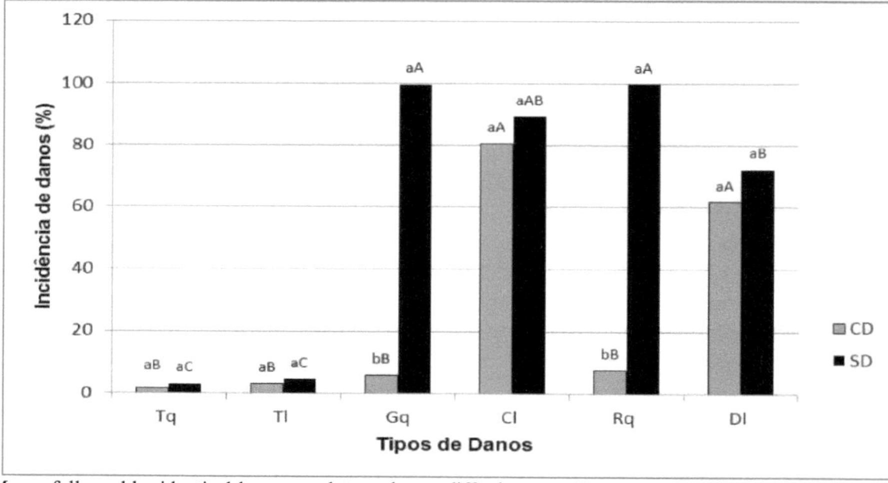

Means followed by identical lowercase letters do not differ between Treatments, and by identical uppercase letters do not differ between Damage Types (Tukey test $p<0.05$).
Note: Tq = breakage of the main stem or trunk; Tl = damage to the trunk reaching the wood; Gq = breakage of secondary branches; Cl = damage to the bark, without reaching the cambium; Rq = foraging of leaves or branchlets; and, Dl = damage greater than 5 cm in diameter.
SOURCE: The Author (2017).

According to Porfirio-da-Silva et al. (2012), branching (Rq) should be expected in silvipastoral systems, especially when animals can reach the branches, suggesting that pruning reduces this occurrence, as well as branch breakage (Gq). Although the branching of some leaves does not have a negative impact on tree development, broken branches can cause injuries that allow pathogenic organisms to enter.

There were significant differences in damage intensity between the treatments with pruning (CD) and without pruning (SD) on *E. benthamii* (Figure 6). In the treatment with pruning (CD), the damage intensity with the highest incidence was d1, with 80.1%, considered an intensity of low importance, while in the treatment without pruning (SD), the damage intensity with the highest incidence was d2, occurring in 84.4% of the trees, an intensity of medium importance for the future development of the tree.

The bark performs a wide range of functions in trees. The innermost layer of the bark is involved in the transportation and storage of photoassimilates, while the outermost layer protects the plant against the entry of pathogens and protects adjacent tissues from mechanical damage (ROMERO, 2006).

The intensity of the damage depends on the depth and extent of the damage to the bark. When the damage is restricted to the surface, without reaching the cambial tissue of the tree, the damaged part recovers evenly. In the case of deeper damage, healing is often only partial, with a portion of the wood remaining exposed, which can serve as a gateway for disease infection and pest attack, even causing the death of trees (MEDRADO et al. 2009). When deep lesions occur in the bark, a carbohydrate derived from parenchyma tissues is deposited to close the lesion (ROMERO, 2006). This can increase cattle attraction to the tree (PORFIRIO-DA-SILVA et al. 2012). Another negative aspect of deep bark damage is the possibility of epicormic sprouts appearing below the damaged area, reducing the value of the wood (PORFIRIO-DASILVA et al., 2012). Only 16.2% of the trees in the pruning treatment (CD) showed no damage (d0). The other trees suffered some kind of damage (Figure 6).

FIGURE 6 - INCIDENCE AND INTENSITY OF DAMAGE CAUSED BY BOVINE ON *E. benthamii* IN THE SILVIPASTORIL SYSTEM WITH DISRUPTION (CD) AND WITHOUT DISRUPTION (SD), UFPR FARM, PINHAIS-PR, 2016.

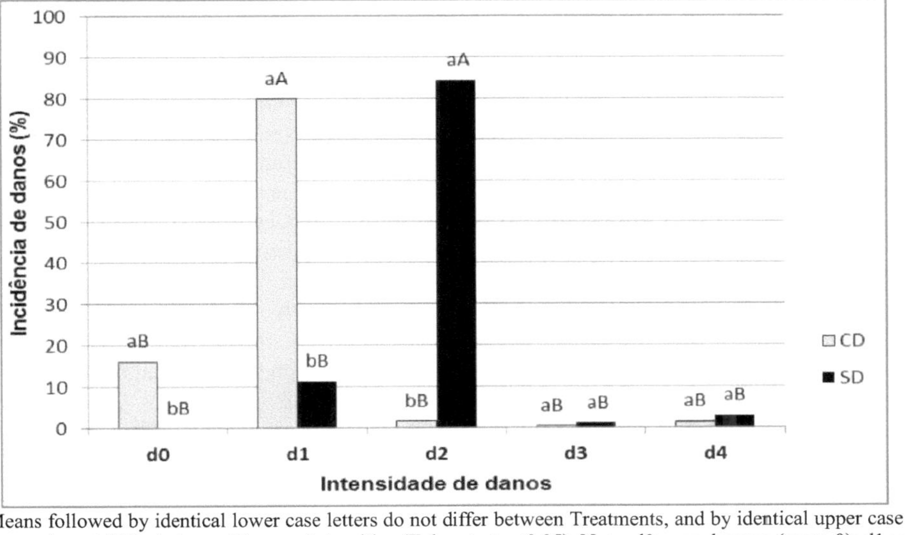

Means followed by identical lower case letters do not differ between Treatments, and by identical upper case letters do not differ between Damage Intensities (Tukey test $p<0.05$). Note: d0 = no damage (score 0); d1 = low damage (scores between 0 and 3); d2 = medium damage (scores between 3 and 6); d3 = high damage (scores between 6 and 10); and, d4 = extreme damage (score 10).

SOURCE: The Author (2017).

Although all the treatments suffered damage from the cattle, there was a significant difference using the Tukey test ($p<0.05$) in the average scores given to the different types of damage. The treatment with pruning (CD) had the lowest average damage score of 1.67, which qualifies it as low damage intensity; while the treatment without pruning (SD), with an average score of 5.03, qualifies as medium damage intensity (Table 2).

TABLE 2 - DAMAGE CAUSED BY BOVINE ON *E.benthamii* TREES WITH SHREDDING (CD) AND WITHOUT SHREDDING (SD) IN A SILVIPASTORIL SYSTEM AND THE RESPECTIVE DIMENSIONS DIAMETER AT BREAST HEIGHT (DBH) AND HEIGHT (H) AT 30 MONTHS, UFPR FARM, PINHAIS-PR, 2016.

Treatment	Damage[(1)]	DBH (cm)	Height (m)
With Pruning (CD)	1,67 b	11,83 a	10,40 a
No Pruning (SD)	5,03 a	11,17 a	9,72 a

[(1)] Average scores for the different types of damage observed.
Means followed by identical letters in the same column do not differ statistically by the Tukey test ($p<0.05$).
SOURCE: The Author (2017).

According to Pinkard (2002), the practice of pruning can reduce the supply of carbohydrates and growth regulators produced in the canopy and therefore reduce the growth in diameter and height of the trees. This effect was reported by Barbosa et al. (2014), when evaluating the effect of the impact of sheep on trees in a silvopastoral system with a hybrid of *E grandis* x *E urophylla,* where at 24 months, they observed that artificial pruning affected tree growth. In the present experiment, there was no such difference in the growth of the DBH and height of the *E. benthamii* trees between the treatments without pruning (SD) and with pruning (CD) (Table 2).

At 30 months, the average DBH of *E. benthamii* was 11.83 cm in the treatment with pruning (CD) and 11.17 cm in the treatment without pruning (SD). The total height was 10.40 m in the treatment with pruning (CD) and 9.72 m in the treatment without pruning (SD). There were no significant differences in DBH and total height between the treatments with pruning (CD) and without pruning (SD).

CONCLUSION

Cattle entering the silvopastoral system 26 months after planting *Eucalyptus benthamii* caused low to medium intensity damage to the trees. Damage to the bark, without reaching the cambium; foraging on leaves or branches and damage greater than 5 cm in diameter were the main damage caused by cattle.

The practice of pruning *Eucalyptus benthamii trees* did not affect the growth in DBH and total height of the trees and contributed to reducing the incidence and intensity of damage caused by cattle.

CHAPTER 3
early growth of *eucalyptus benthamii* Maiden et Cambage IN INTEGRATED AGRICULTURAL PRODUCTION SYSTEMS - SIPA, IN AN ENVIRONMENTAL PRESERVATION AREA

SUMMARY

Trees can improve the productivity of an integrated production system by influencing soil characteristics, microclimate, hydrology and associated biological components. There is still little information on the most suitable tree species for use in integrated agricultural production systems (SIPA), especially in environmental protection areas (APA). Integrated systems have been increasingly adopted in Brazil, even supported by public policies such as the federal government's "Low Carbon Agriculture" program. These systems integrate diversification, income generation and environmental protection. In this context, the aim of this work was to evaluate the growth of *Eucalyptus benthamii* Maiden et Cambage in an APA area, from planting to 36 months of age in SIPA. The experiment was conducted in a randomized block design with four treatments: livestock-forest (LF), crop-livestock-forest (LPF), crop-forest (LF) and monoculture (FM) and three replications, in Pinhais-PR. The perennial summer pasture used in the livestock treatments was *Megathyrsus maximus* - cv Aries with direct sowing of black oats (*Avena strigosa* Schreb.) in winter without desiccation. Systematic inventories were carried out with 10% sampling intensity, at 12 months, 24 months and 36 months, to measure diameter at breast height (DBH) and total height (H). There was no significant variation ($p < 0.05$) between treatments in the total height growth of *Eucalyptus benthamii* at 12 months. At 24 and 36 months, there was a significant variation ($p < 0.05$) between the treatments, with FM outperforming the other treatments. At 36 months, the average total height of the FM treatment was 14.00 m, followed by LF (12.37 m), LPF (11.01 m) and PF (10.88 m). When comparing average DBH growth between treatments, there was a significant difference ($p < 0.05$) only at 12 months of age, with the highest for LF (3.02 cm) and the lowest for FM (2.14 cm). In the evaluations at 24 and 36 months, there were no significant differences ($p < 0.05$) between the treatments. The treatments were not directly related to the number of trees remaining at 36 months. There was no significant variation in the volume of wood per hectare between the PF, LPF and LF treatments at 36 months. However, the volume of wood per hectare of the FM treatment was higher than the other treatments only because of the higher density of plants per hectare.

Keywords: Crop, livestock and forest integration, Agroforestry systems, Eucalyptus, Eucalyptus growth.

INTRODUCTION

Increasing productivity in a sustainable way is the biggest challenge facing the agricultural sector. With an estimated 30% increase in the world's population, demand for agricultural products is expected to rise by approximately 60% globally and 77% in developing countries (OECD/FAO, 2012). To meet this global need, Integrated Agricultural Production Systems (IAPPS) have come to the fore in recent years.

According to Bernardi et al. (2014), the demand for products such as cellulose, resin, wood, among others, has been increasing every year, and SIPAs present themselves as important and viable alternatives to meet these demands, generating additional income. However, maximum productivity of the components of these systems should not be expected, but rather the sustainability of the ecosystem and the generation of satisfactory returns and income from different products.

Interest in silvopastoral and agrosilvopastoral systems has been increasing significantly throughout the country. Practices for integrating crops, livestock and forestry are already appearing in research and extension programs run by government agencies, such as the federal government's "Low Carbon Agriculture" program. These initiatives are the result of the recognition and need to conduct such systems in favor of the soil, the environment and human generations in the present and future (BERNARDINO & GARCIA, 2009; MAPA/ACS, 2012).

Integrating animals with agricultural and forestry crops is not a new system of farming. What is perhaps new is the fact that integrating animals into forestry activities is capable of improving productivity per unit area. Due to the growing increase in the world's population, especially in less developed countries, increasing productivity is becoming very important, since the area available for agriculture will reach its limits in the not too distant future (GARCIA et al., 2003).

The initiative to integrate animals into forestry activities is not a recent technique. In the western United States, the practice of placing cattle to graze in native forest areas has been carried out for over 125 years (ROATH & KRUEGER, 1982). Work involving cattle in pine forests (*Pinus radiata*) in New Zealand was described by Miller (1975). In European countries, too, multiple silvopastoral farming has been carried out for many years. In Brazil, the intentional association of cattle with forests dates back to the mid-18th century in the region where *Araucaria angustifolia* occurs (CHANG, 1985). In a literature review, Porfírio-da-Silva (2015) reports that research into silvipastoral systems in Paraná began in the 1980s with cattle being introduced into conventional forest plantations.

A great deal of research has been carried out into the use of the understorey as a way for wild and domestic animals (mainly cattle and sheep) to make use of it, as well as controlling the development of herbaceous vegetation, which in accumulation becomes a vehicle for spreading fires in forests (SCHREINER, 1983; BAGGIO & SCHREINER, 1988; SCHREINER, 1994; GARCIA et al., 2003) and the compatibility of the associated crop (BERNARDINO, 2007).

Research results prove the potential of SIPA as an important strategy for sustainable rural development. The introduction of a forest component on rural properties through these systems not only ensures more suitable conditions for agricultural crops, pastures and livestock, but also makes it possible to diversify products in the same area (LAURA et al., 2015, PORFÍRIO-DA-SILVA, 2015; STRASSBURG et al., 2014).

Research into integrated production systems in southern Brazil, as in other regions of the country, has increased considerably over the last 20 years. However, the level of adoption of the technologies generated is still considered to be low (DIAS-FILHO & FERREIRA, 2007), despite the enormous potential of these systems to resolve issues of social, economic and environmental importance, as well as meeting demands for the supply of forest products. This is a challenge for the agricultural innovation system in Brazil, i.e. seeking to develop environmentally sustainable agriculture.

The tree component in a production system has great potential for recovering degraded areas, contributing to carbon sequestration, increasing and conserving biodiversity, improving the chemical and physical aspects of the soil, increasing the quality of water and air resources, as well as diversifying the producer's income (SILVA, 2013; JOSE, 2009; QUINKENSTEIN et al., 2009; NAIR, 2007)

Eucalyptus spp. is the most widely planted genus in Brazil, accounting for 72.0% (5.47 million hectares) of the planted forest area. Its main use is in the pulp industry (40.96%), followed by industrial firewood (30.26%) and coal (17.02%). Sawn and treated wood and other solid products account for 6.29% of eucalyptus wood consumption in Brazil (IBA, 2014).

Among the *eucalyptus* species is *Eucalyptus benthamii* Maiden et Cambage, which is recommended for colder regions such as southern Brazil because it has good frost tolerance and normal growth even at average low temperatures such as -1°C to 3°C (PALUDZYSZYN FILHO et al., 2006). It is a species with good suitability for energy purposes (PALUDZYSZYN FILHO & SANTOS, 2011), and its wood is moderately hard at 0.47 g.cm-3 (NISGOSKI et al., 1998), requiring genetic improvement research to identify genotypes suitable for sawmills (SILVA, 2008).

Higa and Carvalho (1990) reported that *E. benthamii* showed 70% survival, an average height of 16 m and an average DBH of 15 cm at 45 months of age in Dois Vizinhos - PR, and concluded that the species deserves special attention from breeders.

Serpe (2015) observed that *Eucalyptus benthamii was* resistant to 15 annual frosts with an absolute minimum temperature of - 6.4 C.°

Silva (2008) concluded that frost resistance is not genetically correlated with growth.

There are studies evaluating the performance of *eucalyptus* species in different arrangements and different managements in SIPA (CALIL et al., 2013; VIEIRA & SCHUMACHER, 2011; KLEINPAUL et al., 2010; MACEDO et al., 2006), but the number of studies on *Eucalyptus benthamii is* still scarce, and there is a need for more in-depth studies on the behavior of this species, in order to better understand it and indicate its management in crop and/or livestock integration systems.

The aim of this study was to evaluate the growth, production characteristics and performance of *Eucalyptus benthamii in* its first three years of age, in monoculture (FM) and in three types of integrated systems: livestock-forest (PF); crop-livestock-forest (LPF) and crop-forest (LF).

MATERIALS AND METHODS

The experiment is part of a project developed by the Agricultural Technology Innovation Center (NITA), located at the Canguiri Experimental Farm (25°23'30"S; 49°07'30" W) belonging to the

University of São Paulo.

Federal University of Paraná, located in the municipality of Pinhais-PR, with an average altitude of 920m. The farm is located within the Irai River Environmental Protection Area (APA) (State Decree No. 1.735/96), on the banks of the Iraí reservoir.

The region's climate, according to Koppen, is classified as Cfb - temperate, humid and with a temperate summer, characterized by an average annual rainfall of 1400 mm, an average minimum temperature of 12.5°C and an average maximum temperature of 22.5°C, being subject to severe frosts (more than five per year).

The average data and curves for rainfall and average temperature indicators for the period 2013-2016 in the experiment area are shown in Figure 7.

According to soil mapping, the classes Cambissolo Háplico, Latossolo Vermelho and Latossolo Vermelho- Amarelo are found in this area (SUGAMOSTO, 2002).

The statistical design was randomized blocks with 3 replications and 4 treatments: T1 - livestock-forest system (PF); T2 - crop-livestock-forest system (LPF); T3 - crop-forest system (LF), and T4 - monoculture forest (FM). Each plot has a different area, with a minimum of 1.3 hectares and a maximum of 2.2 hectares in the PF, LPF and LF treatments and 0.3 to 0.5 hectares in the FM treatment (Figure 8).

FIGURE 7 - RAINFALL AND AVERAGE TEMPERATURE IN THE EXPERIMENTAL AREA, 2013-2016

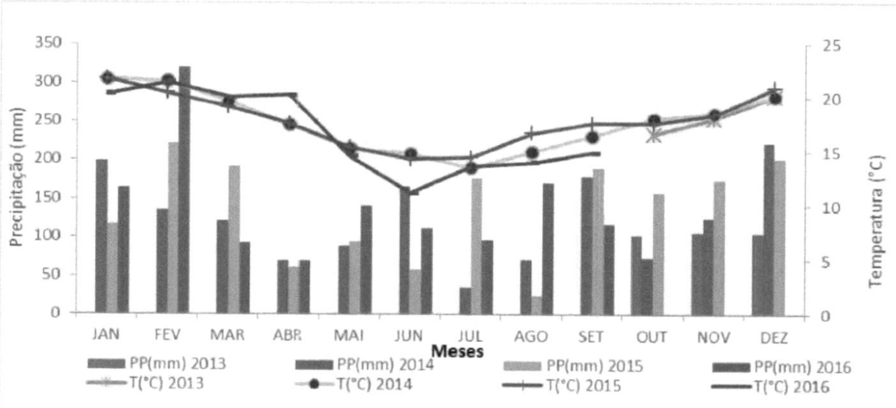

SOURCE: SIMEPAR (2016), adapted by the author.

FIGURE 8 - DISTRIBUTION OF *E. benthamii* IN THE EXPERIMENTAL AREA OF CANGUIRI FARM - UFPR, 2016.

SOURCE: The author, adapted from *Google Earth*.

In the PF, LPF and LF treatments, the trees were planted in single rows with a spacing of 2 m x 14 m, in order to obtain an initial density of approximately 357 trees ha^{-1} and a final density of 150 trees ha^{-1} . The rows of trees were planted on contour lines in order to conserve soil and water, control erosion, improve thermal comfort for the animals and soften the trails made by animals walking on the slope of the land (PORFÍRIO-DA- SILVA et al., 2010). The monoculture forest treatment was planted at a spacing of 2 m x 3 m (1,667 treesha).$^{-1}$

In January and February 2013, the soil was conventionally prepared for sowing perennial summer pasture with *Megathyrsus maximus* - cv Aries. *Megathyrsus maximus* was sown in rows (16 $kg.ha^{-1}$) and 350 $kg.ha^{-1}$ of simple superphosphate was applied in the row and a further 800 $kg.ha^{-1}$ of ammonium chloride. During the establishment of the pasture, three mowings were carried out to reduce the competitive pressure of the weeds.

In September 2013, the tree lines were demarcated and the seedlings of *Eucalyptus benthamii were planted* in October 2013. This species was chosen mainly because it is frost-resistant and the aim is to obtain logs for sawmills and rolling mills at the end of the 20-year cycle.

The eucalyptus trees were planted in the first half of October 2013, and all the treatments received the same soil preparation and tree fertilization. The planting line was subsoiled to a depth of approximately 40 cm, and the planting was done by hand using a weeder to open the holes. For each seedling, 250 ml of Hidroplan-EB®/HyB hydrogel solution was placed in the hole, diluted according to the manufacturer's recommendations, and 200 g of the 08-20-20 NPK formulation, which also contained 1% calcium (Ca) and 3% sulphur (S), was distributed in two lateral pits (15 cm away from the seedling and 15 cm deep).

Leaf-cutting ants were controlled by applying Formifu® paste to all the seedlings fifteen days after planting. Orange peels were also used and distributed weekly at strategic points in and around the area to attract the ants and locate their nests, which were then destroyed mechanically.

In the process of planting the trees, it was necessary to replant 20% of the trees in the experiment (January 2014). As this is an APA, the use of herbicides is not permitted, and weed control was carried out manually with a hoe and motorized with a three-stroke brush cutter.

Maize and sunflower were sown in the spring-summer no-till system in the crop-forest treatment (LF) at the end of September and beginning of October, without the use of herbicides or any other type of agrochemical due to the fact that the experiment was located in an Environmental Protection Area (APA). The spacing of the two crops was 0.45 x 0.20 m (111,111 $plants.ha^{-1}$). In the autumn-winter period, black oats (*Avena strigosa* Schreb.) were sown in a no-till system at the end of May and beginning of June, in order to keep the soil covered with vegetation during the period.

In the PF and LPF treatments, with the perennial summer pasture (*Megathyrsus maximus* - cv Aries), in the autumn-winter period, black oats (*Avena strigosa* Schreb.) were over-seeded in late May and early June, in a no-till system, to provide pasture in winter or to keep the soil covered with vegetation during the period in the LPF treatment, when cultivated with a summer crop.

The fertilization of the production system was standardized and determined by NITA's technical team, based on soil analysis and the demands of the most demanding crop (maize).

In the first year (2013), the fertilizers were applied in two stages: the first in September, using 450 $kg.ha^{-1}$ of simple superphosphate and 400 $kg.ha^{-1}$ of the NPK 25:00:25 formula, the second in October, applying 200 $kg.ha^{-1}$ of urea.

[a]In the second year (2014), the fertilizers were applied to the treatments in three stages: the first after planting the winter pasture (oats), in the 2nd half of May, using 600 $kg.ha^{-1}$ of natural phosphate, 200 $kg.ha^{-1}$ potassium chloride and 200 $kg.ha^{-1}$ of urea; the second when the annual summer crops (corn and sunflower) were planted in September, using 450 $kg.ha^{-1}$ of simple superphosphate and the third in November, with 450 $kg.ha^{-1}$ of urea.

In 2015, the fertilizers were applied in two stages: the first, in the 2nd half of May after planting the winter pasture (oats), using 300 $kg.ha^{-1}$ of natural phosphate, 100 $kg.ha^{-1}$ potassium chloride and 300 $kg.ha^{-1}$ of urea; the second applying 200 $kg.ha^{-1}$ of urea at the end of December.

In 2016 (the fourth year), after planting the winter pasture (oats), 300 $kg.ha^{-1}$ of natural phosphate, 100 $kg.ha^{-1}$ potassium chloride and 300 $kg.ha^{-1}$ of urea were applied to all treatments in the second half of May.

The trees were first pruned at 21 months of age (July 2015) to half their respective heights (on average up to 3 m), using pruning saws. The 2nd pruning was carried out at 33 months of age (July 2016), cutting the branches to half the trees' respective trunk heights (up to 5 to 6 m). In the FM treatment, natural pruning took place during the 2nd pruning with the death of the lateral branches, requiring only the pruning of the trees at the edge of the plots.

The animals entered the experiment in the PF and LPF treatments at 24 months of tree age (October 2015), when the trees had an average DBH of 8.12 cm, which is considered adequate to support their presence (MEDRADO et al. 2009; PORFIRIO-DASILVA et al. 2012). The areas of the LP and LPF treatments were divided into paddocks, covering 1.3 to 2.2 hectares each (Figure 8), using electric fencing. In all the paddocks, the animals were provided with water and mineralized salt at will.

At the end of May and beginning of June 2016, in the PF and LPF treatments, with the perennial summer pasture (*Megathyrsus maximus* - cv), black oats (*Avena strigosa* Schreb.) were over-seeded in a no-till system, for grazing in the winter-spring period.

Tree growth was monitored by collecting data from the trees in each treatment (PF, LPF, LF and FM), using systematic sampling of 10% of the trees, measuring the diameter at breast height (DBH) at 1.30 m above ground level and total height (H). The DBH and total height of all the trees were measured at 12, 24 and 36 months. At 12 months, the DBH was measured with a *King.Tools®* digital caliper by taking two perpendicular measurements on each tree and calculating the average. To obtain the DBH for the other measurements, the circumference at breast height (CBH) was measured with a tape measure and the circumference was transformed into a diameter (CBH=BCH/n). To measure the total height of the trees in meters, a 4 m graduated ruler was used at 12 months and a Haglof® electronic clinometer at 24 and 36 months.

The volume per tree sampled was obtained using the expression:

$$V = \frac{DAP^{2} . \pi . H. f}{40000} \quad (1)$$

Being:

Volume = volume of the tree with bark in cubic meters

DBH = diameter at breast height in centimeters (1.30 m from the ground) H = total tree height in meters *f* = 0.45 (shape factor)*[1]

The volume per hectare was obtained by multiplying the volume of the average tree, i.e. the tree with the estimated average DBH and total height of the treatment population, by the number of trees remaining per hectare in each period, for each treatment evaluated. The average annual increment (IMA) in volume per hectare (m^3 ha^{-1}) was calculated by dividing the total volume per hectare by the current age of the forest stand, in years, at the time of each assessment. The annual current increment (ICA) of volume per hectare (m^3 ha^{-1}) was calculated by dividing the total volume per hectare in the year of each assessment.

The percentage of trees remaining in the field was calculated for all treatments by subtracting the number of trees from the initial planting of each plot from the number of trees existing at each evaluation. As this was a plantation intended to produce wood for sawmills, dead or broken trees were excluded, as well as trees with dry tips, crooked trees, fallen trees and other deformities. There was also a need to exclude some trees near the fences dividing the plots so that machinery could pass through.

The homogeneity of variance between treatments was analyzed using Bartlett's test to verify the normality of the data as a prerequisite for the use of analysis of variance (ANOVA) and the means were compared using Tukey's test at a 5% probability level to evaluate the growth of DBH and Height within treatment and between treatments. Only the remaining trees were considered for the analysis. The statistical analyses were carried out using the ASSISTAT statistical program version 7.7 en (2016), at 5% probability of error (SILVA, 2009).

RESULTS AND DISCUSSION

When comparing the treatments, there was a significant difference ($p < 0.05$) in the DBH growth of *Eucalyptus benthamii* only at 12 months of age. The treatment with the highest DBH growth at 12 months was LF with 3.02 cm and the one with the lowest growth was FM with 2.14 cm, which has a higher planting density. These values corroborate those of Bernardo (1995) and Oliveira Neto et al. (2003), who found that the effect of increasing density was to reduce DBH growth, which was expressed in the early stages of plant development. At 24 and 36 months, the DBH did not differ significantly ($p < 0.05$) between the treatments (Table 3). At 36 months, the average DBH of the LF treatment was 15.08 cm, followed, in descending order, by the PF (DBH = 13.87 cm), LPF (DBH = 13.20 cm) and FM (DBH = 12.45 cm) treatments.

TABLE 3 - BREAST HEIGHT DIAMETER (DBH) AND TOTAL HEIGHT OF *E. benthamii* IN

*[1] Average used for rapid estimation of the total volume of standing trees with monopodial growth, Porfírio-da-Silva et al. (2010).

MONOCULTIVATION (FM) AND INTEGRATED SYSTEM WITH CROP AND PASTURE (LF, PF and LPF) AT 12, 24 and 36 MONTHS, UFPR FARM, PINHAIS-PR, 2016.

Immediately	Diameter at breast height-DAP (cm)			Total height (m)		
	12 months	24 months	36 months	12 months	24 months	36 months
PF	2.41 ab	8,12 a	13,87 a	2,52 a	6,74 b	10,88 b
LPF	2.41 ab	8,13 a	13,20 a	2,61 a	7,05 b	11,01 b
LF	3,02 a	9,97 a	15,08 a	3,04 a	7.90 ab	12,37 b
FM	2,14 b	8,80 a	12,45 a	2,56 a	9,12 a	14,00 a

PF=Livestock-Forest; LPF= Crop-Livestock-Forest; LF= Crop-Forest; FM= Forest-Massif.
Means followed by identical letters in the same column do not differ statistically by the Tukey test ($p<0.05$).
SOURCE: The Author (2017).

Although it did not differ significantly ($p > 0.05$) between the treatments, the lowest average value at 24 months found for the DBH of the FM treatment at 8.80 cm is equivalent to 88.26% of the LF treatment (DBH = 9.97 cm), which was the highest DBH value among the treatments for that age. This percentage dropped to 82.56% at 36 months. This indicates that the growth rate in DBH decreases over time in plantations with higher densities, as verified by Leite et al. (1997).

Benin *et al.* (2014), obtained for *E. benthamii* in the third year after planting, an average annual increase in DBH of 2.85, 3.10, 3.24 and 3.43 cm, respectively at initial spacings of 2 m x 3 m, 3 m x 3 m, 4 m x 3 m and 4 m x 4m. Pereira *et al.* (2000) at four years of age, with an initial spacing of 3 m x 3 m, observed a DBH of 14.2 cm (3.55 cm per year^{-1}), and at eight years of age it grew by 2.7 cm per year^{-1} . Higa & Carvalho (1990), on the other hand, obtained an annual diameter growth of 3.75 cm per year for *E. benthamii* at 4 years of age^{-1} and 18.2 cm per year (2.28 cm per year^{-1})

at eight years of age [HIGA & PEREIRA, 2003]. Serpe (2015), using a spacing of 2.5 m x 2.5 m, observed that at 6 years of age the DBH varied from 11.65 cm in the treatment without fertilizer to 15.25 cm in the best result. Souza (2015) found that at 24 months of age, DBH ranged from 5.2 cm in the treatment without fertilizer to 8.0 cm in the best result, very close to the values found in this experiment.

At 36 months, the DBH of the different treatments ranged from 12.45 to 15.08 cm. When compared to the average diameter growth of the aforementioned studies, diameter growth was above the average found in the literature. This performance can partly be attributed to the appropriate cultural management of fertilization in this experiment and the good adaptation of *E. benthamii to the* soil and climatic conditions of the site and the treatments used.

With regard to the total height growth of *E. benthamii*, at 12 months there was no significant variation ($p < 0.05$) between the treatments (Table 3). At 24 and 36 months, there was a significant variation ($p < 0.05$) between the treatments, with FM superior to the other treatments. At 36 months, the average total height of the FM treatment was 14.00 m, followed by LF (12.37 m), LPF (11.01 m) and PF (10.88 m), which had the lowest total height.

The behavior of the *E. benthamii* trees in terms of height growth does not reflect the growth in diameter. At 36 months of age, the FM treatment had the greatest total height (14.00 m). As for DBH, although there was no significant difference ($p < 0.05$) between the treatments, the highest value observed at 36 months of age was in the LF treatment, with 15.08 cm, while in the FM treatment it was 12.50 cm.

Although this experiment showed greater height in the FM treatment at 24 and 36 months, compared to the other treatments with lower densities, Bernardo (1995) mentions that the average height of trees decreases over time as the density of stands increases, due to the increase in dominated trees. In general, diameter growth is a characteristic that is highly responsive to spacing. However, there is some controversy as to the effects on tree height in the young growth phase, with some cases in which there is an increase in height with greater spacing and others in which the result is the opposite (OLIVEIRA et al., 2009).

Bernardi (2010) observed a height growth of 5.64 m for *E. benthamii* in monoculture at 12 months of age. Souza (2015), at 24 months, observed a height ranging from 6.6 m in the treatment without fertilizer to 9.2 m in the best result, compatible with the values found in this study, i.e. a range from 6.74 to 9.12 m (Table 3). Higa & Carvalho (1990) obtained an average height for *E. benthamii* at 4 years of age of 16 m (4 $m.year^{-1}$). At eight years of age, Pereira *et al.* (2000) observed an average height growth of 2.3 m per $year^{-1}$. Higa & Pereira (2003), on the other hand, observed an average height of 21.7 m (2.71 m per $year^{-1}$) at eight years of age. Serpe (2015) observed a height growth of 3.2 m per year at 6 years of age^{-1} .

When compared to the average growth in height, it can be seen that in the present experiment, *E. benthamii,* at 36 months, showed growth in height in the treatments ranging from 3.63 to 4.67 m per $year^{-1}$, which is above the average found in most of the literature. This performance can partly be attributed to the appropriate cultural management of fertilization in this experiment and the good adaptation of *E. benthamii to* the soil and climatic conditions of the site.

The treatment with the best growth performance in terms of DBH in the three periods evaluated (12, 24 and 36 months) was LF. In terms of total height, at 12 months the LF treatment had the best growth performance and at 24 and 36 months the FM treatment had the best growth.

There were significant differences between the treatments and the three evaluation periods for the following variables: remnancy, volume per hectare, average annual increment (IMA) and current annual increment (ICA) (tables 4, 5 and 6). The exception was the variable volume per tree, where analysis of the results showed that there was no significant variation between the treatments.

At 12 months, the LF treatment had the lowest remnant rate (78.0%) and the PF treatment had the highest remnant rate (93.5%) (Table 4). At 36 months, there was no significant variation ($p < 0.05$) in survival between the treatments, but the LF treatment continued to have the lowest numerical value (75.5%) and the highest survival rate was in the LPF treatment (82.0%). The initial survival of the *E. benthamii* seedlings was affected by competition with weeds, especially the viola rope (*Ipomoea sp*), the incidence of which in some plots reached more than 50% of the seedlings. As this is an APA area, where the use of herbicides is not permitted, weed control was carried out manually with a hoe and grubbing, and motorized with a backhoe three times in the first year, but several seedlings ended up being damaged.

TABLE 4 - REMANESCENCE AND FINAL DENSITY OF *E. benthamii* IN MONOCULTURE (FM) AND INTEGRATED WITH CROP AND PASTURE (PF, LPF and LF) AT 12, 24 and 36 MONTHS, FAZENDA UFPR, PINHAIS-PR, 2016.

Immediately -	Remaining rate (%)			Final density (plants.ha)$^{-1}$		
	12 months	24 months	36 months	12 months	24 months	36 months
PF	93.5 aA	90.2 aA	79.8 aB	334	322	285
LPF	91.5 abA	89.8 aA	82.0 aB	327	320	293
LF	78.0 bA	78.0 aA	75.5 aA	279	279	270
FM	88.1 abA	86.7 aA	80.5 aB	1468	1445	1342

AVERAGE	87,8	86,1	79,5	602	591	547

PF=Livestock-Forest; LPF= Crop-Livestock-Forest; LF= Crop-Forest; FM= Forest-Massif.
Means followed by identical lowercase letters in the same column and identical uppercase letters in the same row do not differ statistically by the Tukey test ($p<0.05$).
SOURCE: The Author (2017).

In the three evaluation periods, the treatments did not influence the productive performance of the individual plants, considering that there was no significant difference in the volume per tree (Table 5). This variable is directly related to height and DBH (PEREIRA et al., 2015). In this study, the highest DBH value at 36 months occurred in the LF treatment and the lowest in the FM treatment, reflecting the result of the LF treatment, which had the highest volume per plant (0.0995 m^3), 29.56% higher than the FM treatment (0.0768 m).[3]

At four and eight years of age, *E. benthamii had an* individual commercial volume of 0.1110 and 0.2850 m^3 (PEREIRA *et al.*, 2000). Serpe (2015) observed that at 6 years of age the average individual volume varied from 0.1335 m3 without fertilization to 0.2276 m3 in the best treatment with fertilization. In the present study, at 36 months the average individual volume in the different treatments ranged from 0.0679 to 0.0995 m3, compatible with the values found in the literature.

Volume production per hectare was influenced by the number of trees in the treatments, so that the FM treatment with the highest initial density (1,667 trees ha^{-1}) achieved the highest volume per hectare in the three seasons evaluated, reaching 103.51 m^3 ha^{-1} at 36 months. In the other treatments, with the lowest initial density (357 trees ha^{-1}), at 36 months the volume ranged from 21.23 to 29.84 m^3 ha^{-1} (table 5). In other words, in this study, volume production per hectare was directly proportional to the density of the stand in the treatments, with the highest values at the end of the evaluation period (36 months) in the FM treatment (6.0 m2 $plant^{-1}$) and the lowest values in the PF, LPF and LF treatments (28.0 m2 plant).$^{-1}$

Botelho (1998) and Oliveira Neto et al. (2003), mention that there is greater production per unit area in smaller spacings due to the greater number of individuals. Macedo et al. (2010) observed that the higher volume per hectare in denser spacings was also observed by other authors, stating that the variable volume per hectare is closely related to the number of trees per hectare. In this experiment, up to 36 months, the FM treatment, which is the densest among the treatments (1,667 trees ha^{-1}), was the one that achieved the highest wood production (103.51 m^3 ha^{-1}), confirming this statement.

TABLE 5 - INDIVIDUAL VOLUME PER TREE AND TOTAL VOLUME PER HECTARE OF *E. benthamii IN* MONOCULTIVATION (FM) AND INTEGRATED WITH CROP AND PASTURE (PF, LPF and LF) AT 12, 24 and 36 MONTHS, UFPR FARM, PINHAIS-PR, 2016.

Treatment	Volume per tree (m3)	Volume per hectare (m3)

	12 months	24 months	36 months	12 months	24 months	36 months
PF	0,0005	0,0157	0,0740	0.18 aB	5.57 bB	22.50 bA
LPF	0,0005	0,0165	0,0679	0.18 aB	5.64 bAB	21.23 bA
LF	0,0010	0,0278	0,0995	0.29 aB	9.00 bB	29.84 bA
FM	0,0004	0,0250	0,0768	0.62 aC	36.71 aB	103.51 aA

PF=Livestock-Forest; LPF= Crop-Livestock-Forest; LF= Crop-Forest; FM= Forest-Massif.
For volume per tree, the comparison of means test was not applied because the F interaction was not significant.
Means followed by identical lowercase letters in the same column and identical uppercase letters in the same row do not differ statistically by the Tukey test ($p<0.05$).
SOURCE: The Author (2017).

Serpe (2015) found that at 6 years of age the total volume varied from 180.2 m3 ha^{-1} , in the treatment without fertilization, to 326.2 m3 ha^{-1} in the best treatment with fertilization.

At 12 months, there was no significant variation (p <0 .05) in the average annual increment (IMA) and current annual increment (ICA) between treatments (Table 6). At 24 months and 36 months, there was a significant variation between treatments and evaluation periods. At 36 months, the treatment with the highest IMA value was FM with 34.50 m^3 ha^{-1} , while the other treatments ranged from 7.07 to 9.95 m^3 ha^{-1} . The ICA of the FM treatment at 36 months was 22.27 m^3 ha^{-1} , and varied from 5.20 to 6.95 m^3 ha^{-1} in the other treatments (Table 6).

TABLE 6 - AVERAGE ANNUAL GROWTH (AMI) AND CURRENT ANNUAL GROWTH (CAH) OF *E. benthamii* IN MONOCULTIVATION (FM) AND INTEGRATED WITH CROP AND PASTURE (PF, LPF and LF) AT 12, 24 and 36 MONTHS, UFPR FARM, PINHAIS-PR, 2016.

Immediately -	Average Annual Increase (m^3 .ha $)^{-1}$			Annual Current Increase (m^3 .ha $)^{-1}$		
	12 months	24 months	36 months	12 months	24 months	36 months
PF	0.06 aB	1.36 bB	7.50 bA	0.06 aB	1.80 bAB	5.64 bA

LPF	0.06 aB	1.88 bAB	7.07 bA	0.06 aB	1.82 bAB	5.20 bA
LF	0.10 aB	3.00 bB	9.95 bA	0.10 aB	2.90 bB	6.95 bA
FM	0.21 aC	12.24 aB	34.50 aA	0.21 aC	12.03 aB	22.27 aA

PF=Livestock-Forest; LPF=Livestock-Forest; LF=Livestock-Forest; FM=Forest-Massif.
Means followed by identical lowercase letters in the same column and identical uppercase letters in the same row do not differ statistically by the Tukey test ($p<0.05$).
SOURCE: The Author (2017).

Benin *et al.* (2014) obtained a volume per hectare of 31.61, 34.72, 41.05 and 66.55 m^3 ha^{-1} in the third year, respectively at initial spacings of 4 m x 4 m, 4 m x 3 m, 3 m x 3 m and 2 m x 3 m. Serpe (2015), at 6 years of age, found an IMA ranging from 30.00 m3 ha^{-1} $year^{-1}$, in the treatment without fertilization, to 54.40 m3 ha^{-1} $year^{-1}$ in the treatment with fertilization. Dias (2016), for *Eucalyptus benthamii,* recorded average ICA and IMA for the different treatments of 52.40 and 32.31 m3 ha $year^{-1}$ in the third year, while in the fourth year ICA and IMA were 137.95 and 58.72 †‡ *2 m^3 ha $year^{-1}$, respectively.

Dias (2016) in an experiment studying phosphate fertilization with different doses in the municipality of Otacílio Costa (SC), with a 2.5 x 2.5 m spacing, for *Eucalyptus benthamii,* recorded average ICA and IMA of the different treatments of 52.40 and 32.31 m3 ha $year^{-1}$ in the third year, and in the fourth year ICA and IMA of 137.95 and 58.72*[3] m3 ha $year^{-1}$ were recorded, respectively.

CONCLUSIONS

At 36 months, the average growth in DBH and total height of *Eucalyptus benthamii* in the PF, LPF, LF and FM treatments was compatible with that reported in monocultural experiments. The treatments were not directly related to the number of trees remaining at 36 months. There was no significant variation in the volume of wood per hectare between the PF, LPF and LF treatments at 36 months. However, the volume of wood per hectare in the FM treatment was higher than in the other treatments only because of the higher density of plants per hectare.

---

*[2] Values without shape factor applied to volume calculation

‡[3] Values without the shape factor applied to the volume calculation

FINAL CONSIDERATIONS

The aim of this study was to provide information on the tree component (eucalyptus) in integrated agricultural production systems (SIPA) by evaluating its performance in initial establishment in the environmental protection area (APA) of the Iraí River, in Pinhais/Paraná, Brazil.

It was found that it is possible to establish the tree component in SIPA without using pesticides.

The world's predominant agricultural production model is simplistic from an ecological point of view and has significant potential to damage the environment and profitability in the production of grains, fibers and animal protein, compromising the sustainability of agricultural systems. The need to build diversified, more heterogeneous production systems is of fundamental importance for the sustainability of the rational use of natural resources. The use of animals in pastures integrated with agriculture and forestry crops greatly stimulates diversity and the generation of important ecosystem services.

Among the strategies for increasing diversity in agricultural production systems, the introduction of a tree component is a suitable option for small, medium and large farms, both in grain-producing regions and those traditionally producing livestock and forest crops. By increasing the variability of the living organisms that make up the agro-ecosystem and by diversifying economic activities on the farm, the producer's income is diversified due to the greater variety of products (grain, meat or milk and wood) that can be sold throughout the year, making it possible for rural producers to remain in business.

5.1 SUGGESTIONS FOR FUTURE WORK

Although the results of this study are encouraging for the use of *Eucalyptus benthammi* in SIPA in subtropical conditions, there is a need for further experiments, as this is the initial performance of the eucalyptus cycle, which is generally around 18 years.

In addition to the silvicultural performance of *Eucalyptus benthammi* in SIPA until it completes the harvest cycle, there are still issues to be worked on, such as the best time to carry out thinning and harvesting, the ideal density for each integration system, the quality and use of the wood, and the dynamics of nutrient cycling in the forest component. In this sense, in addition to the continuation of existing experiments, new research is also needed.

APPENDIX A - MISCELLANEOUS TABLES

TABLE 7- TEMPERATURES AND RAINFALL IN THE AREA FOR THE PERIOD 2013-2016.

Average Maximum Temperature (°C)												
	JAN	FEB	MAR	APR	MAY	JUN	JUL	AUG	SET	OUT	NOV	TEN
2013										22,7	23,6	26,2
2014	28,7	28,1	25,2	22,6	20,7	20,1	19,2	22,1	22,3	25,4	24,5	26,6
2015	28,6	26,7	24,6	23,5	20,3	20,3	19,9	24,0	24,0	23,2	23,6	26,5
2016	26,0	27,4	25,9	27,3	19,6	17,3	20,7	20,6	22,1			
Average Minimum Temperature (°C)												
	JAN	FEB	MAR	APR	MAY	JUN	JUL	AUG	SET	OUT	NOV	TEN
2013										12,5	14,2	16,2
2014	17,4	17,3	16,1	14,5	11,8	11,1	9,5	9,2	12,5	13,4	14,8	16,1
2015	17,6	16,9	16,3	13,9	12,2	9,5	10,3	11,4	12,9	14,1	15,4	17,5
2016	16,8	18,1	16,5	15,8	11,2	7,0	8,3	8,9	9,9			
Temperature M							**average (°C)**					
	JAN	FEB	MAR	APR	MAY	JUN	JUL	AUG	SET	OUT	NOV	TEN

	JAN	FEB	MAR	APR	MAY	JUN	JUL	AUG	SET	OUT	NOV	TEN
2013										16,6	18,0	20,1
2014	21,8	21,6	19,7	17,5	15,3	14,8	13,5	15,0	16,4	18,0	18,5	20,1
2015	21,8	20,5	19,2	17,6	15,4	14,3	14,5	16,7	17,6	17,6	18,4	20,9
2016	20,4	21,5	20,1	20,3	14,6	11,2	13,7	14,0	14,9			
Absolute Maximum Temperature (°C)												
	JAN	FEB	MAR	APR	MAY	JUN	JUL	AUG	SET	OUT	NOV	TEN
2013										29,5	31,1	32,0
2014	33,0	34,6	30,2	28,8	26,8	24,7	27,0	28,5	28,7	34,0	31,0	31,1
2015	33,8	30,9	28,3	28,7	26,3	25,7	25,9	30,5	33,9	32,0	32,3	31,1
2016	31,5	31,9	31,2	31,8	24,2	24,9	27,7	28,8	29,5			
Absolute Minimum Temperature (°C)												
	JAN	FEB	MAR	APR	MAY	JUN	JUL	AUG	SET	OUT	NOV	TEN
2013										7,1	9,3	13,1
2014	14,6	13,6	11,9	9,9	8,6	1,3	2,4	3,9	8,7	9,6	9,1	11,5
2015	14,6	14,6	13,1	9,8	6,0	1,0	6,0	7,4	6,3	9,1	12,8	13,6

2016	1,6	15,1	13,2	2,4	2,2	-1,8	0,3	0,8	4,7			
Precipitation (mm)												
	JAN	FEB	MAR	APR	MAY	JUN	JUL	AUG	SET	OUT	NOV	TEN
2013										101,6	105,6	103,8
2014	198,6	135,2	121,2	68,4	88,0	164,8	34,4	69,8	178,0	72,6	124,0	220,6
2015	117,4	221,0	191,0	60,8	93,8	56,6	175,6	24,4	189,4	157,4	173,2	200,4
2016	164,2	319,6	92,2	68,4	140,8	111,0	96,4	170,6	116,6			

SOURCE: SIMEPAR (2016)

TABLE 8 - DATA ON THE DEPTH OF THE A HORIZON AND CHEMICAL ANALYSIS OF THE SOIL (0 - 20 CM) IN THE AREA OF THE EXPERIMENT ON THE UFPR FARM, PINHAIS, 2013.

Treatment	Horizon A	pH		Al^{+3}	H+Al +3	Ca^{+2}	Mg^{+2}	K+3	SB	T	P	CV	M	Ca/mg
	(cm)	CaCl2	SMP			C i	molc/ d	Im^3			Mg/dm^3	g/dm $\%^3$	%	
PF-Bl 1	95,00	5,47	6,00	0,00	5,00	6,20	3,63	0,22	10,05	15,05	2,73	29,27 66,67	0,00	1,70
PF-Bl 2	100,00	5,47	5,83	0,00	5,67	6,63	3,87	0,15	10,65	16,32	2,27	20,57 65,00	0,00	1,71
PF-Bl 3	35,00	5,90	6,37	0,00	3,83	4,80	3,13	0,08	8,01	11,84	1,17	13,03 67,33	0,00	1,53
LPF-Bl 1	88,33	5,60	6,03	0,03	4,93	6,47	3,70	0,23	10,40	15,33	3,00	33,50 68,00	0,33	1,75

LPF-Bl 2	82,67	5,50	5,90	0,00	5,40	6,57	3,80	0,23	10,60	16,00	3,07	71,03 66,67	0,00	1,72
LPF Bl 3	53,33	5,73	6,20	0,00	4,37	5,63	3,17	0,07	8,87	13,24	4,23	20,60 67,00	0,00	1,78
LF-Bl 1	81,67	4,97	5,77	0,17	6,00	5,03	2,87	0,10	8,00	14,00	2,83	29,70 64,33	0,00	1,71
LF-Bl 2	75,00	5,53	5,97	0,17	5,73	6,20	3,60	0,17	9,97	15,70	2,60	27,77 63,67	1,67	1,73
LF-Bl 3	50,00	5,60	6,10	0,00	4,60	5,60	3,70	0,15	9,45	14,05	2,30	27,40 67,00	0,00	1,51
FM-Bl 1	97,50	5,65	6,25	0,00	4,20	5,30	3,50	0,12	8,92	13,12	0,50	21,80 68,00	0,00	1,52
FM-Bl 2	75,67	5,57	5,93	0,00	5,87	6,37	3,73	0,18	10,28	16,14	2,83	29,70 64,33	0,00	1,71
FM-Bl 3	91,67	5,50	5,90	0,00	5,50	6,20	4,00	0,06	10,26	15,76	1,23	22,57 65,33	0,00	1,58

PF=Livestock-Forest; LPF=Livestock-Forest; LF=Livestock-Forest; FM=Forest-Macizo; Bl= Block.
SOURCE: The Author (2017).

APPENDIX B - PHOTOS

Figure 1: Images of the E. benthamii planting at UFPR's Canguiri Experimental Farm, Pinhais-PR, October 2013. A: planting line following the contour of the land; B: preparing the hole with hydrogel solution; C: seedling after transplanting; D: general aspect of the planting operation.
SOURCE: The Author (2017).

Figure 2: Images of the treatments in April 2014. A: treatment LF - Crop and Forest; B: treatment LPF - Crop, Livestock and Forest; C: treatment PF - Livestock and Forest; D: treatment FM - Massive Forest. UFPR Canguiri Experimental Farm, Pinhais-PR.
SOURCE: The Author (2017).

Image of the general aspect of the experiment 01 year after it was set up (October 2014). UFPR's Canguiri Experimental Farm, Pinhais-PR.
SOURCE: The Author (2017).

Figure 4: Measurement images 01 year (October 2014). A: general view of the measurement work; B: measurement of the total height of the trees using a topographic sight; C: B: measurement of the breast height - DBH of the trees using a digital caliper. UFPR Canguiri Experimental Farm, Pinhais-PR.
SOURCE: The Author (2017).

Figure 5 Images 1ª pruning (July 2015). A: detail of pruning with a pruning saw; B: general appearance of the trees after pruning; Canguiri Experimental Farm at UFPR, Pinhais-PR.
SOURCE: The Author (2017).

Figure 6 Images of the 2nd pruning (July 2016). A: pruning treatment LPF - Crop, Livestock and Forest; B: pruning treatment LF - Crop and Forest. UFPR's Canguiri Experimental Farm, Pinhais-PR.
SOURCE: The Author (2017).

Figure 7 Images of cattle entering the experiment (October 2015). A: LPF treatment - Crop, Livestock and Forest; B: pruning PF treatment - Livestock and Forest. UFPR's Canguiri Experimental Farm, Pinhais-PR.
SOURCE: The Author (2017).

Figure 8. Images of the treatments in November 2016. A: treatment PF- Livestock and Forest; B: treatment LPF- Crop, Livestock and Forest (crop and forest phase); C: treatment LF- Crop and Forest; D: treatment FM- Massive Forest. Canguiri Experimental Farm at UFPR, Pinhais-PR.
SOURCE: The Author (2017).

Figure 9. Images of cattle damage to trees in the PF treatment (silvipastoral) during spring/summer 2015/2016 A: Tq damage = breakage of the main stem or trunk in a tree with a DBH of less than 6 cm; B: Tl damage = damage to the trunk reaching the wood by removal of cambial tissue; C: Dl damage = damage greater than 5 cm in diameter; D: Gq damage = breakage of secondary branches; E: Cl damage = damage to the bark, without reaching the cambium; F: Rq damage = breakage of thin branches and foraging of leaves, or ramoneio. UFPR Canguiri Experimental Farm, Pinhais-PR.
SOURCE: The Author (2017).

REFERENCES

ABRAF - BRAZILIAN ASSOCIATION OF PRODUCERS OF PLANTED FORESTS. Statistical yearbook. ABRAF 2013, base year 2012 / ABRAF. Brasília: 148 p. 2013.

ACOSTA, M. S. Solid products from cultivated eucalyptus wood. Entre Ríos: INTA, (CIDEU, Boletín). 15 p. 2006.

ALMEIDA, R.G.; ZIMMER, A.H.; KICHEL, A.N.; MACEDO, M.C.M.; COSTA, J.A.A. Pasture recovery strategies through integrated beef production systems. In: CONGRESSO BRASILEIRO DE ZOOTECNIA - ZOOTEC, 22.2012, Cuiabá. Proceedings. Cuiabá: UFMT; ABZ, 2012. p. 1-34. 1 CD-ROM.

ALVES, I. C. N.; GOMIDE, J. L.; COLODETTE, J. L.; SILVA, H. D. da. Technological characterization of *Eucalyptus benthamii* wood for Kraft pulp production. Ciência Florestal, Santa Maria, v. 21, n. 1, p. 167-174, 2011.

AMPARADO, K. F.; CARVALHO, A. M.; GARCIA, R. A.; LATORRACA, J. V. F. Characterization of *Eucalyptus saligna* Smith sawn timber yield in green and dry conditions. Revista Florestal Venezolana, ano XLII, v. 52, p. 71-76, 2008.

ANDRADE, E. N. O Eucalipto. Sao Paulo: Cia. Paulista de Estradas de Ferro. 2ª Edition, 680p. 1961.

ANGELI, A. Indications for choosing Eucalyptus species. Supervised by BARRICHELO, L. E. G. and MULLER, P. H. IPEF. 2005. http://www.ipef.br/identificacao/eucalyptus/indicacoes.asp. Accessed on: Nov. 18, 2016.

ANGHINONI, I.; CARVALHO, P. C. F.; COSTA, S. E. V. G. A. Topics in Soil Science. In: Araújo, A. P.; Avelar, B. J. R., (Eds.) Abordagem sistêmica do solo em sistemas integrados de producao agrícola e pecuária no subtrópico brasileiro. 8th ed. Viçosa: UFV, chap. 8, p. 221-278. 2013.

AYRES, E. C. B.; RIBEIRO, A. E. M. Agroecological innovations in the northeast of Minas Gerais: the case of agroforestry systems in family farming in the upper Jequitinhonha region. Organizaçoes Rurais & Agroindustriais, Lavras, v. 12, n. 3, p. 344-354, 2010.

BAGGIO, A. J.; SCHREINER, H. G. Analysis of a silvopastoral system with *Pinus elliottii* and beef cattle. Boletim de Pesquisa Florestal, Curitiba, n. 16, p. 19-29, 1988.

BALBINO, L. C.; CORDEIRO, L. A. M.; PORFÍRIO-DA-SILVA, V.; MORAES, A; MARTÍNEZ, G. B.; ALVARENGA, R. C.; KICHEL, A. N.; FONTANELI, R. S.; SANTOS, H. P.; FRANCHINI, J. C.; GALERANI, P. R. Technological evolution and productive arrangements of crop-livestock-forest integration systems in Brazil. Pesquisa Agropecuária Brasileira, Brasília, v.46, n.10, p. 1-12, 2011.

BARCELLOS, D. C.; COUTO, L. C.; MULLER, M. D.; COUTO, L. The state-of-the-art of eucalyptus wood quality for energy production: a focus on silvicultural treatments. Biomassa e Energia, Viposa, v. 2, n. 2, p. 141-158, 2005.

BENSON, D. H. Aspects of the ecology of a rare tree species, *Eucalyptus benthamii*, at Bents Basin, Wallacia. Cunnighamia, v. 1, n. 3, p. 371-383, 1985.

BENSON, D.; McDOUGALL, L. Ecology of Sydney plant species: part 6 dicotyledon family Myrtaceae. Cunninghamia, Sydney, v. 5, n. 4, p. 809987, 1998.

BERNARDI, C. A. Evaluation of different soil preparation systems on the initial development of *Eucalyptus benthamii* Maiden et Cambage in the region of Guarapuava, PR. Irati, 46p. Dissertation (Master's Degree in Forestry Sciences) - Midwestern State University, 2010.

BERNARDI, Camila Motta Marin; MACEDO, H. R., PINHEIRO, R. S. B., & FREITAS, M. L. M. Eucalyptus planted forests in silvipastoral systems and the impact of animal input. Revista Verde de Agroecologia e Desenvolvimento Sustentável, v. 9, n. 5, p. 125-132, 2014.

BERNARDINO, F. S. Silvipastoral system with eucalyptus: understory productivity and performance of steers under nitrogen and potassium fertilization. Viposa: UFV, 112 p. PhD thesis. 2007.

BERNARDINO; F. S.; GARCIA; R. Sistemas silvipastoris. Research Florestal Brasileira, Colombo, n.60, p.77-87, Special Edition. dec. 2009.

BOCAGE, I.; ULERY, A. Anatomical characterization of *Eucalyptus benthamii* Maiden et Cambage grown in Uruguay. In: jornada forestal: visita a ensayos de silvicultura y mejoramiento de pinus y eucaliptos. Colonvade. 2005.

BRAZIL. Law No. 6.902, of April 27, 1981. Provides for the creation of Ecological Stations,

Environmental Protection Areas and other measures. Office of the President of the Republic. Brasilia: 1981.

BRAZIL. Law No. 12.805, of April 29, 2013. Institutes the National Crop-Livestock-Forest Integration Policy and amends Law No.° 8.171, of January 17, 1991. Office of the President of the Republic. Brasilia: DOU de 30.4.2013. 2013.

BRAZIL. Ministry of Agriculture, Livestock and Supply. Sectoral plan for mitigation and adaptation to climate change for the consolidation of a low carbon economy in agriculture: ABC plan (Low Carbon Agriculture). Ministry of Agriculture, Livestock and Supply, Ministry of Agrarian Development, coordination of the Civil House of the Presidency of the Republic. Brasília: MAPA/ACS, 173 p. 2012.

BRONDANI, G. E.; WENDLING, I.; DUTRA, L. F.; GROSSI, F. Vegetative Propagation of E. *benthamii* x *E. dunnii* by Miniestaquia. Colombo: Embrapa Florestas, 42p. (Embrapa Florestas. Documentos, 183). 2009.

BUTCHER, P. A.; SKINNER, A. K.; GARDINER, C. A. Increased inbreeding and inter-species gene flow in remnant populations of the rare *Eucalyptus benthamii*. Conservation Genetics, v. 6, n. 2, p. 213-226, 2005.

CAIXETA, R. P.; TRUGILHO, P. F.; ROSADO, S. C. S.; LIMA, J. T. Wood properties and classification applied to the selection of Eucalyptus genotypes. Árvore, Viposa, v. 27, n. 1, p. 23-51, 2003.

CARVALHO, M. M.; ALVIM, M. J.; XAVIER, D. F. et al. Establishment of silvopastoral systems: emphasis on mountainous areas and low fertility soils Anais do ZOOTEC'2005 - May 24-27, 2005 - Campo Grande-MS. Juiz de Fora: Embrapa Gado de Leite (Technical Circular, 68). 2002.

CARVALHO, M. M.; FREITAS, V. P.; ANDRADE, A.C. Initial growth of five tropical grasses in an angico-vermelho (Anadenanthera macrocarpa Benth.) understory. Past. Trop., v.17, p.24-30, 1995.

CARVALHO, P. C. de F.; MORAES, A. de, SILVEIRA PONTES, L. da, ANGHINONI, I., SULC, R. M., BATELLO, C. Definipoes e terminologias para Sistema Integrado de Produpao Agropecuária. Revista Ciencia Agronómica, v. 45, n. 5, p. 1040-1046, 2014.

CASTRO, C.R.T.; PACIULLO, D.S.C. Good practices for implementing silvopastoral systems. Juiz de Fora: Embrapa Gado de Leite, 6 p. (Embrapa Gado de Leite. Comunicado Técnico, 50). 2006.

CHANG, M.Y. Faxinais in Paraná. Londrina, PR: Instituto Agronómico do Paraná - IAPAR, 26p. 1985.

CORDEIRO, L. A. M., VILELA, L., MARCHÄO, R. L., KLUTHCOUSKI, J., MARTHA-JÚNIOR, G. B. Crop-Livestock Integration and Crop-Livestock-Forest Integration: strategies for sustainable intensification of land use. Cadernos de Ciencia & Tecnología, v. 32, n. 1/2, p. 15-53, 2016.

DA SILVA, L. A. C., & DE LIMA MEIRELLES, P. R. Use of eucalyptus in integrated animal production systems. *JORNACITEC, 2*. 2013.

DANIEL, O.; COUTO, L. Diagnosis of the Situation and Needs for the Development of SAFs in Mato Grosso do Sul. In: Seminar "Agroforestry systems and sustainable development", 2003, Campo Grande. Proceedings... Campo Grande: Embrapa Gado de Corte15 p. 1 CDROM, 2001.

DARROW, W. K. Selection of eucalypt species for cold and dry areas in South Africa. In: CRCTHF-IUFRO CONFERENCE, 1995, Hobart. Eucalypt plantations: improving fiber yield and quality. Hobart: CRC, 1995. p. 336338. Available at: <http://www.forestry.crc.org.au/iufro95.htm#theme2>. Accessed on: June 14, 2005.

DE ALMEIDA, R. G.; MACEDO, M. C. M.; ALVES, F. V. Integración de sistemas de cultivo-ganadería-forestal con énfasis en la producción de carne. Universidad Nacional de Colombia, Medellín, II Congreso Colombiano y 1er Seminario Internacional Silvopastoreo, 2012.

DIAS-FILHO, M. B. Silvipastoral systems in the recovery of degraded tropical pastures. In.GONZAGA NETO, S.; COSTA, R. G.; PIMENTA FILHO, E. C.; CASTRO, J. M. C. Simpósio a reunião anual da sociedade brasileira de zootecnia, 43. João Pessoa. Proceedings... João Pessoa: SBZ: UFPB, 2006 (Special supplement to Revista Brasileira de Zootecnia, v. 35). p. 535-553. 2006.

EMBRAPA. Adoption of ILPF reaches 11.5 million hectares. Available at : <https://www.embrapa.br/busca-de-noticias/-

/noticia/17755008/adocao-de-ilpf-chega-a-115-milhoes-de- hectares?link=agencia>. Accessed on: 04 Nov. 2016.
Brazilian Agricultural Research Corporation - EMBRAPA. National Forest Research Center, Curitiba, PR. Ecological zoning for forest plantations in the state of Santa Catarina, by Antonio Aparecido Carpanezzi and others. Curitiba, EMBRAPA CNPF, 113 p. (EMBRAPA-CNPF. Documentos, 21). 1988.
FAO - *Food and Agriculture Organization Of The United Nations*. Eucalyptus in forest repopulation. Rome: Food and Agriculture Organization of the United Nations, p. 723. 1981.
FAO, FLD, IPGRI. Forest genetic resources conservation and management. Rome: International Plant Genetic Resources Institute, 94 p. 2004.
FARREL, J. G.; ALTIERI, M. Agroforestry systems. In: Altieri, M. Agroecologia: bases científicas para uma agricultura sustentável. 3 Ed. Sao Paulo, Rio de Janeiro: Expressao Popular. AS- APTA. p. 281-304, 2012.
FOELKEL, C. E. B.; BARRICHELO, L. E. G.; MILANEZ, A. F. Comparative study of Eucalyptus saligna, E. paniculata, E. citriodora, E. maculata and E. tereticornis wood for the production of sulphate cellulose. IPEF, Piracicaba, n. 10, p. 17-37, 1975.
GALZERANO, L.; MORGADO, E. Eucalyptus in agroforestry systems. Revista Electronica de Veterinária, Garpa-SP v. 9, n. 3, 2008.
GARCIA, R.; COUTO, L.; ANDRADE, C. M. S.; TSUKAMOTO FILHO, A. de A. Silvipastoral systems in the southeast: CMM's experience. Text of the lecture presented at the Seminar "Agroforestry Systems and Sustainable Development" - Campo Grande MS, 2003. http://saf.cnpgc.embrapa.br/publicacoes/22.pdf>. Accessed Nov. 18, 2016.
GLASER, F.D., 2008. Behavioral aspects of Angus, Caracú and Nelore cattle grazing on the availability of shade and water for immersion. 117 f. Thesis (Doctorate in Animal Science) - University of Sao Paulo (USP), Pirassununga. 2008.
GRAQA, M. E. C.; SHIMIZU, J. Y.; TAVARES, F. R. Regrowth and rooting capacity of *Eucalyptus benthamii*. Boletim de Pesquisa Florestal, Colombo, n. 39, p. 135-138, 1999.
HIGA, A. R.; CARVALHO, P. E. R. Survival and growth of twelve eucalyptus species in Dois Vizinhos, Paraná. Silvicultura, Sao Paulo, v. 2, n. 42, p. 459-461, 1990.
HIGA, R. C. V. Ecological and silvicultural aspects of *Eucalyptus benthamii* Maiden et Cambage. Boletim de Pesquisa Florestal; 38: 121123. 1999.
HIGA, R. C. V.; MORA, A. L.; HIGA, A. R. Planting eucalyptus on small rural properties. Colombo: Embrapa Florestas, 31p. (Embrapa Florestas. Documentos, 54). 2000a
HIGA, R. C. V.; PEREIRA, J. C. D. Potential Uses of *Eucalyptus benthamii* Maiden et Cambage. Colombo: Embrapa Florestas, (Embrapa Florestas. Comunicado Técnico, 100). 2003.
HIGA, R. C. V.; WREGE, M. S.; HIGA, A. R.; SANTOS, G. A. Considerations on the climatic requirements of *Eucalyptus benthamii*. In: SILVA, Luciana Duque; HIGA, Antonio Rioyei; SANTOS, Glêison Augusto dos (Coord.). Silviculture and genetic improvement of *Eucalyptus benthamii*. Curitiba, PR: Fupef, Chap. 1. p. 11-19. 2012.
HIGASHI, E. N.; SILVEIRA, R. L. V. de A.; GONÇALVES, A. N. Vegetative propagation of *Eucalyptus*: basic principles and its evolution in Brazil. Instituto de Pesquisas e Estudos Florestais - IPEF, 11 p. (IPEF Technical Circular, 192). 2000.
IBÁ (BRAZILIAN TREE INDUSTRY). Ibá Annual Report 2014. Brasília: 100 p., 2014.
IBÁ (BRAZILIAN TREE INDUSTRY). Ibá Annual Report 2016. Brasília: 100 p., 2016.
BRAZILIAN INSTITUTE OF GEOGRAPHY AND STATISTICS - IBGE. Municipal livestock survey. 2006. Available at: http:<www. ibge.gov.br>.Acesso em: 18 nov 2016.
JAEGER, P.; ZIGER, M. Evaluation of the mechanical properties of panels JANKOWSKY, I. P.; SANTOS, G. R. V.; ANDRADE, A. Drying of eucalyptus lumber. In: Jornadas Florestales de Entre Rios, 15, 2000, Concordia. Proceedings... Concordia: [S. I.], p. C2-1-C2-13. 2000.
JOVANOVIC, T.; BOOTH, T. H. Improved species climatic profiles. Australia: Union Offset Printing: Joint Venture Agroforestry Program, Rural Industries Research and Development Corporation, 68 p. 2002.
LIMA, E. A. de; SILVA, H. D da; LAVORANTI, O. J. Dendroenergetic characterization of *Eucalyptus benthamii* trees. Pesq. flor. bras., Colombo, v. 31, n. 65, p. 09-17, 2011.
LIMA, E. A. de; SILVA, H. D da; TUSSOLINI, E. L. Potential of *Eucalyptus benthamii* for charcoal production in conventional kilns. Colombo: Embrapa Florestas, 4p. (Embrapa Florestas. Technical Communication, 305). 2012.

LIMA, I. L. Influence of thinning and fertilization on the quality of sawn wood from Eucalyptus grandis Hill ex-Maiden. 137 f. Thesis (PhD in Forest Resources) - University of Sao Paulo, Piracicaba, 2005.
LOBAO, S. M.; LUCIA, R. M. D.; MOREIRA, M. S. S.; GOMES, A. Caracterizapao das propriedades físico-mecánicas da madeira de eucalipto com diferentes densidades. Árvore, Viposa, v. 28, n. 6, p. 889-894, 2004.
MACEDO, R.L.G.; VALE, A.B.; VENTURIN, N. Eucalyptus in agroforestry systems. Lavras: UFLA, 331p. 2010.
MARCHIORI, J. N. C.; SOBRAL, M. Dendrology of Angiosperms: Myrtales. Santa Maria: Editora UFSM, 304 p. 1997.
MARTHA JUNIOR, G. B.; VILELA, L. Land-saving effect of crop-livestock integration systems. Planaltina, DF: Embrapa Cerrados, 2009. 4 p. (Embrapa Cerrados. Technical Communication, 164). Available at < http://bbeletronica.cpac.embrapa.br/versaomodelo/html/2009/comtec/comt ec_164.shtml>. Accessed on: April 5, 2017.
MARTIN, G. O. Keep the shade on your foals and reduce animal stress. Revista Producción. 2002. Available at <http://www.ecampo.com/sections/news/print.php/uuid.582F356F-2996- 417A-8D93D1A411F549BD>Accessed: Nov. 18, 2016.
MARTINS, J. L.; SILVA, I. J. O. S.; FAGNANI, M. A. et al. Evaluation of the thermal quality of the natural shading of some tree species in pasture conditions. In: SBZ. Annual Meeting of the Brazilian Society of Animal Science, 39, Recife. CD-ROM. SBZ, 2002
MEDRADO, M. J. S. M.; SILVA, V. P. da; DERETI, R. M., FONSECA, L.R. da; MAIER, T. F.; PINTON, A. L. M. Damage caused to eucalyptus by cattle raised in a silvipastoral system in the municipality of Cruzmaltina, PR. Technical Communication 243. ISSN 1517-5030. Colombo, PR. 2009.
MELOTTO, A.M.; LAURA, V.A.; BUNGENSTAB, D. J. The forest component in integrated production systems. In: BUNGENSTAB, D. J. (Ed.). Crop-livestock-forest integration systems: sustainable production. Campo Grande: Embrapa Gado de Corte, p. 37-51. 2011.
MENDOZA, L. Notes on *Eucalyptus benthamii* in Argentina. In: COLLOQUES INTERNATIONAL SUR LES EUCALYPTUS RESISTANTS AU FROID, 1983, Bordeaux. Annales... Bordeaux: IUFRO, p. 480. 1983.
MILLER, T. Cattle and Trees thriving together. New Zealand Journal Agricultural, p. 53-55, 1975.
MIRANDA, M. D. A.; NAHUZ, M. A. R. Study of the influence of *Eucalyptus saligna* plantation spacing on splitting rates after splitting and after drying. Scientia Florestalis, n. 55, p. 107-116, 1999.
MORA, A. L.; GARCIA, C. H. The cultivation of eucalyptus in Brazil. Sao Paulo: SBS, 2000.
MORAES, A.; CARVALHO, P. C. F.; BARRO, R. S.; LUSTOSA, S. B. C.; PORFÍRIO-DA-SILVA, V.; REISENDORF-LANG, C. Prospects for research into integrated crop and livestock production systems in Brazil and the new challenges. In: ANAIS ANNUAL MEETING OF THE BRAZILIAN SOCIETY OF ZOOTECHNY. Brasilia, DF, 2012.
MORAES NETO, S. P.; PULROLNIK, K.; VILELA, L.; MUNHOZ, D. J. de M.; GUIMARÄES JÚNIOR, R.; MARCHÄO, R. L. Hypsometric models for *Eucalyptus cloeziana* and *Eucalyptus uruphilla* x *Eucalyptus grandis* in an Agroforestry system. Planaltina-DF: Embrapa Cerrados. Research and development bulletin, 286, 2010.
MOTT, G. O.; LUCAS, H. L. The design, conduct and interpretation of grazing trials on cultivated and improved pastures. In: INTERNATIONAL GRASSLAND CONGRESS, 6, 1952, Pennsylvania. Proceedings..., Pennsylvania, p. 1380 - 1385. 1952.
MOURA, V. P. G.; GUIMARAES, D. P. Production of Eucalyptus seedlings for the establishment of forest plantations. Brasilia: Embrapa, 9 p. (Embrapa, Technical Communication, 85). 2003.
MÜLLER, B.V. Effect of splitting systems on the quality and yield of sawn timber from *Eucalyptus benthamii* Maiden et Cambage. Curitiba, 119p. Dissertation (Master's Degree in Forestry Engineering) - Federal University of Paraná, 2013.
NAIR, P.K.R. An Introduction to Agroforestry. Kluwer Academic Publishers, Dordrecht, The

Netherlands. 499 p. 1993.
NAIR, P. K. R. Carbon sequestration studies in agroforestry systems. Agroforest Systems. v.86, n.2, p.243-253, 2012.
NICODEMO, M.L.F.; SILVA, V.P.; THIAGO, L.R.L.S.; LAURA, V.A. Sistemas silvipastoris: introdupao de árvores na pecuária do Centro-Oeste brasileiro. Campo Grande: Embrapa Gado de Corte, 37 p. (Embrapa Gado de Corte. Documentos, 146). 2004.
NISGOSKI, S.; MUÑIZ, G. I. B. de; KLOCK, U. Anatomical characterization of *Eucalyptus benthamii* Maiden et Cambage wood. Ciencia Florestal, Santa Maria, v. 8, n. 1, p. 67-76, 1998.
OECD/FAO (*ORGANIZATION FOR ECONOMIC COOPERATION AND DEVELOPMENT / FOOD AND AGRICULTURE ORGANIZATION OF THE UNITED NATIONS*). OECD-FAO Agricultural Outlook 2012-2021. OECD Publishing and FAO. 2012.
OLIVEIRA NETO, S. N. de; VALE, A. B. do; NACIF, A. de P.; VILAR, M. B.; ASSIS, J. B. de. Agrosilvipastoral system: crop, livestock and forest. Viposa-MG, Sociedade de Investigapóes Florestais, 190 p. 2010.
OLIVEIRA NETO, S. N., PAIVA, H.N. Implementation and management of the tree component in an agroforestry system. In: Oliveira Neto, S. N., Vale, A.B., Nacif, A.P., Vilar, M.B., Assis, J.B. (Ed.). Agroforestry system: integrating crops, livestock and forests. Viposa: Sociedade de Investigapóes Florestais, p.15-68, 2010.
UN, World population to reach 9.6 billion by 2050. Available at: http://www.onu.org.br/populacao-mundialdeve-atingir-96-bilhoes-em-2050- diz-novo-relatorioda-onu/. 2013. Accessed on: November 18, 2013.
PALUDZYSZYN FILHO, E.; SANTOS, P. E. T. Considerapóes sobre o plantio de *Eucalyptus dunnii* no estado do Paraná. Colombo: Embrapa Florestas, 7p. (Embrapa Florestas. Comunicado Técnico, 141). 2005.
PALUDZYSZYN FILHO, E.; SANTOS, P. E. T. dos; FERREIRA, C. A. Eucalyptus trees suitable for planting in the state of Paraná. Colombo: Embrapa Florestas - CNPF, 45 p. (Documentos, 129). 2006.
PARANÁ. Decree No. 1.753, of May 6, 1996. Establishes the Environmental Protection Area in the source area of the Iraí River basin, called the Irai State APA. Curitiba. Official Gazette No. 4.750. 1996.
PEREIRA, J. C. D.; SCHAÍTZA, E. G; SHIMIZU, J. Physical and chemical characteristics and dry distillation yields of *Eucalyptus benthamii* Maiden et Cambage wood. Colombo: Embrapa Florestas, 4p. (Embrapa Florestas. Technical Circular, 50). 2001.
PEREIRA, J. C. D.; STURION, J. A.; HIGA, A. R.; HIGA, R. C. V.; SHIMIZU, J. Wood characteristics of some eucalyptus species planted in Brazil . Colombo: Embrapa Florestas, 113p. (Embrapa Forestry. Documents, 38). 2000.
PONCE, R. H. Eucalyptus sawnwood: challenges and prospects. In: International Seminar on the Use of Eucalyptus Wood for Sawmills. Sao Paulo, p. 50-58. 1995.
POPAY, I.; FIELD, R. Grazing animals as weed control agents. Weed Technology, v. 10, n.1, p. 217-231, 1996.
PORFÍRIO-DA-SILVA, V. Ecology and management in silvopastoral systems. In: FERNANDES, E.N. *et al.* (Ed.). Agroforestry systems in South America: challenges and potential. Juiz de Fora: Embrapa Gado de Leite, p. 51-67, 2007b.
PORFÍRIO-DA-SILVA, V. MORAES, A. Silvipastoral systems: fundamentals for implementation. In: Pires, A. V. (Ed.). Bovinocultura de corte, Piracicaba: FEALQ, v. 2, p. 1421-1461. 2010.
PORFÍRIO-DA-SILVA, V. Silvipastoral system for meat production. In: SYMPOSIUM ON PASTURE MANAGEMENT, 2006, Piracicaba. Pastures and the environment: proceedings. Piracicaba: FEALQ, p. 297-327. 2006.
PORFÍRIO-DA-SILVA, V.; MEDRADO, M. J. S.; NICODEMO, M. L. F; DERETI, R. M. Afforestation of Pastures with Woody Forest Species: implementation and management. Colombo, Embrapa Florestas, 48 p. 2010.
PORFIRIO-DA-SILVA, V.; MORAES, A. de; MOLETTA, J.L., PONTES, L.S.; OLIVEIRA, E.B. de; PELISSARI, A.; CARVALHO, P.C.F. de. Cattle damage to different tree species recommended for silvopastoral systems. Pesquisa Florestal Brasileira 32:67-76, 2012.
PORFIRIO-DA-SILVA, V. Crop-livestock-forest integration.

Opinioes, ano 12, n. 40, p. 6-7, jun./ago. Special editorial. 2015.
PORFIRIO-DA-SILVA, V. The silvipastoral system in Paraná: a synopsis. *In:* SIMPÓSIO DE PRODUQÃO ANIMAL A PASTO, 3., 2015, Maringá. Proceedings. Maringá: Sthampa, p. 253-272. 2015.
PRYOR, L. D. Australian endangered species: *Eucalyptus*. Canberra: Commonwealth of Australia. 139 p. 1981.
PRYOR, L. D.; JOHNSON, L. A. S. A classification of the eucalypts. Canberra: Australian National University, 1971.
PULROLNIK, K.; VILELA, L.; MORAES NETO, S. P. de; MARCHÃO, R. L.; GUIMARÃES JÚNIOR, R. Initial Development of Tree Species in the Crop-Livestock-Forest Integration System. Planaltina-DF: Embrapa Cerrados, (Research and Development Bulletin - 276). 17p. 2010.
ROATH, R.L., KRUEGER, W.C. Cattle Grazing and Behavior on a Forested Range. Journal of Range Management, Denver, v.35, n.3, p.332338, 1982.
ROCHA, M. P. *Eucalyptus grandis* Hill ex Maiden and *Eucalyptus dunnii* Maiden as Sources of Raw Material for Sawmills. 185 f. Thesis (Doctorate in Forest Engineering) - Agricultural Sciences Sector, Federal University of Paraná, Curitiba, 2000.
RODRIGUES, R. A. D. Variability of physical-mechanical properties in batches of eucalyptus sawn timber for civil construction. 75 f. Dissertation (Master's Degree in Forest Resources) - University of Sao Paulo, Piracicaba, 2002.
ROZAS MELLADO, E. C. E. Contribution to technological development for the use of sawn wood from *Eucalyptus grandis* (Hill Ex Maiden) in the generation of products with greater added value. 133 f. Dissertation (Master's Degree in Forestry Engineering) - Agricultural Sciences Sector, Federal University of Paraná, Curitiba, 1993.
SANTOS, A. F. dos; AUER, C. G.; GRIGOLETTI JÚNIOR, A. Eucalyptus diseases in southern Brazil: identification and control. Colombo: EMBRAPA- CNPF. Technical Circular 45, 20 p. 2001.
SCHREINER, H.G. Feasibility of a silvopastoral system on quartz sand soils in the State of São Paulo. Boletim de Pesquisa Florestal, Colombo, v. 17. p. 33-38. 1988.
SCHREINER, H.G. Technical consultancy report on agroforestry. Contractor: National Forestry Research Center/EMBRAPA. 76p. 1994.
SERPE, Edson Luis. Effects of different fertilizer dosages on the initial growth of *Eucalyptus benthamii* in the southern region of the state of Paraná -- Irati, PR: [s.n], 82f. Dissertation (master's degree) - Postgraduate Program in Forest Sciences, area of concentration in Forest Management. UNICENTRO. 2015.
SILVA, I. C. Agroforestry systems: concepts and methods. 1 ed. - Itabuna: SBSAF. 308 p., 2013.
SILVA, L.L.G.G.; RESENDE, A.S.; DIAS, P.F.; SOUTO, S.M. et al. Brachiaria brizantha cv. Marandu in a silvipastoral system. Seropédica: Embrapa Agrobiologia, 28 p. (Embrapa Agrobiologia. Boletim de Pesquisa e Desenvolvimento, 33). 2008b.
SNIF (NATIONAL FOREST INFORMATION SYSTEM). Planted forests. http://www.florestal.gov.br/snif/recursos-forests/planted-forests. Accessed on: 15 Oct. 2016.
SOUZA, A. N.; OLIVEIRA, A. D.; SCOLFORO, J. R. S.; REZENDE, J. L. P.; MELLO, J. M. Economic viability of an agroforestry system. Cerne, Lavras, v. 13, n. 1, p. 96-106, 2007.
STAHL, J.; LAZZAROTTI, G.; NAMIKAMA, I. S.; CHAVES, D. M.; MOREIRA, A. M.; MENDES, C. J.; PUCCI, J. A. Silviculture of *Eucalyptus benthamii* at Klabin S/A. In: SILVA, L. D.; HIGA, A. R.; SANTOS, G. A. dos S. (Coor.). Silviculture and genetic improvement of *Eucalyptus benthamii*. Curitiba, PR: Fupef, Chap. 6. p. 105-121. 2012.
STRASSBURG, B. B., LATAWIEC, A. E., BARIONI, L. G., NOBRE, C. A., PORFÍRIO-DA-SILVA, V., VALENTIM, J. F. & ASSAD, E. D. When enough should be enough: Improving the use of current agricultural lands could meet production demands and spare natural habitats in Brazil. *Global Environmental Change, 28,* 84-97, 2014.
SUGAMOSTO, M. L. Use of geoprocessing techniques to draw up the agricultural suitability map and evaluate the suitability of use of the Cangüiri experimental station center, municipality of Pinhais - Paraná. 149f. Dissertation (Master's Degree in Agronomy) - Postgraduate Program in Agronomy, Federal University of Paraná. 2002.

TRECENTI, R.; OLIVEIRA, M. C.; HASS, G. Crop-Livestock-Forestry Integration. Brasília: MAPA, (Technical Bulletin). 54 p. 2008.
UTIMA, A.Y. Growth and yield of the agricultural and tree components of an integrated agricultural production system in the year of implementation in an environmental protection area. Curitiba, 73p. Dissertation (Master's Degree in Agronomy and Plant Production) - Federal University of Paraná, 2015.
VENTURIN, R. P.; GUERRA, A. R.; MACEDO, R. L. G.; VENTURIN, N; MESQUITA, H. A. Agrosilvipastoral systems: origin, modalities and implementation models. In: Informe Agropecuário, Belo Horizonte, v. 31, n. 257, p. 16-24. 2010.
VILELA, L; MARTHA-JÚNIOR, G. B.; MARCHÃO, R. L. Crop-livestock-forest integration: an alternative for intensifying land use. Revista UFG/December, p. 92-99, 2012.
VITTI, A. M. S.; BRITO, J. O. Essential oil of eucalyptus. Forestry Documents No. 17, ESALQ/USP. 2003.
ANDERSON, G. W.; HAWKE, M.; MOORE; R. W. Pine needle consumption and bark stripping by sheep grazing annual pastures in young stands of widely spaced Pinus radiata and P. pinaster. Agroforestry Systems 3:37-45, 1985.
BARBOSA; C. M. P.; GONZALEZ, L. R.; CAQÃO, M. M. de F.; BRITO, J. de J.; SILVA FILHO, O. C.; PORFÍRIO-DA-SILVA, V. Damage caused by sheep to eucalyptus trees in a silvipastoral system distributed in two spatial models. *IN*: International Symposium on Pasture Afforestation in Subtropical Regions (1: 2013: Curitiba, PR). Colombo: Embrapa Florestas (Documentos 268), 2014.
BERNARDI, C. M. M.; MACEDO, H. R., PINHEIRO, R. S. B., & FREITAS, M. L. M. Eucalyptus planted forests in silvipastoral systems and the impact of animal input. Revista Verde de Agroecologia e Desenvolvimento Sustentável, v. 9, n. 5, p. 125-132, 2014.
BERNARDINO, F. S. and GARCIA, B. Silvipastoral systems. Pesquisa Florestal Brasileira, Colombo, n.60, p.77-87, dec. Special Issue. 2009.
BERRIDGE, K. C.; KRINGELBACH, M. L. Affective neuroscience of pleasure: reward in humans and animals. Psychopharmacology, v. 199, n. 3, p. 457-480, 2008.
CARVALHO, M. M.; ALVIM, M. J.; XAVIER, D. F. et al. Establishment of silvipastoral systems: emphasis on mountainous areas and low-fertility soils. Juiz de Fora: Embrapa Gado de Leite (Technical Circular, 68). 2002.
DIAS-FILHO, M. B.; FERREIRA, J. N. Barriers to the adoption of silvipastoral systems. In: SIMPÓSIO DE FORRAGICULTURA E PASTAGENS, 6., 2007, Lavras. Theme in evidence: cost-benefit ratio: proceedings. Lavras: NEFOR: UFLA, p. 347-365. 2007.
GARRET, H. E.; KERLEY, M. S.; LADYMAN, K. P.; WALTER, W. D.; GODSEY, L. D.; VAN SAMBEEK, J. W.;, BRAUER, D.K. Hardwood silvopasture management in North America. Agrofor Syst 61:21-33. 2004.
GILL, R. M. A review of damage by mammals in north temperate forests: 3. impact on trees and forests. Forestry. 65:363-388. 1992.
GUERREIRO, M. F.; NICODEMO, M. L. F.; PORFIRIO-DA-SILVA, V. Vulnerability of ten eucalyptus varieties to predation by cattle in a silvopastoral system. Agroforestry System, n. 89, p.743-749, 2015.
AGRONOMIC INSTITUTE OF PARANÁ - IAPAR. Climate maps of the state of Paraná: climate classification. 2000. Available at: <http://www.iapar.br/modules/conteudo/conteudo.php?conteudo=863> Accessed on: 07 Dec. 2016.
LEHMKUHLER J., FELTON E.E.D., SCHMIDT D.A., BADER K.J., MOORE A., HUCK M.B., GARRETT H.E., KERLEY M.S. Cattle performance and tree damage using various tree protection methods during tree establishment. Agrofor Syst 59:35-42. 2003.
LUNELLI, Naiana P.; RAMOS, Soraia F.; JÚNIOR, Clovis JF Oliviera. Agroforestry and externalities. Green Journal of Agroecology and Sustainable Development, v. 8, n. 5, p. 163-170, 2014.
MALAFAIA, P., BARBOSA, J.D., TOKARNIA, C.H. & OLIVEIRA, C.M.C. Behavioral disorders in ruminants not associated with diseases: origin, significance and importance. Pesquisa Veterinária Brasileira, v. 31, n. 9, p. 781-790, 2011.
MASON, G.; CLUBB, R.; LATHAM, N. et al. Why and how should we use environmental enrichment to tackle stereotypic behavior? Applied Animal Behaviour Science 102:163-188,

2007.
MEDRADO, M. J. S. M.; SILVA, V. P. da; DERETI, R. M., FONSECA, L.R. da; MAIER, T. F.; PINTON, A. L. M. Damage caused to eucalyptus by cattle raised in a silvipastoral system in the municipality of Cruzmaltina, PR. Technical Communication 243. ISSN 1517-5030. Colombo, PR. 2009.
MOTT, G. O.; LUCAS, H. L. The design, conduct and interpretation of grazing trials on cultivated and improved pastures. In: INTERNATIONAL GRASSLAND CONGRESS, 6, 1952,
NOLTE DL (2003) Repellents are socially acceptable tools.West For 48:22-23. ttp://digitalcommons.unl.edu/cgi/viewcontent.cgi?article=1256&context=icw dm_usdanwrc. Accessed on: Aug. 15, 2016.
PINKARD, E.A. Effects of pattern and severity of pruning on growth and branch development of pre-canopy closure *Eucaliptus nitens*. Forest Ecology and Management, v. 157, p.127-230, 2002.
PORFIRIO-DA-SILVA V., MORAES A. de, MOLETTA J.L., PONTES L.S. da, OLIVEIRA E.B. de, PELISSARI A., CARVALHO P.C.F. de. Cattle damage to different tree species recommended for silvopastoral systems. Pesquisa Florestal Brasileira, 32:67-76. 2012.
PORFÍRIO-DA-SILVA, V., MEDRADO, M.J.S., NICODEMO, M.L.F., DERETI, R.M. Afforestation of pastures with timber forest species: implementation and management. Colombo: Embrapa Florestas, 48p, 2010.
PORFÍRIO-DA-SILVA, V.; MORAES, A. de; MOLETTA, J. L.; SILVEIRA PONTES, L. da, OLIVEIRA, E. B. de; PELISSARI, A.; CARVALHO, P. C. de F. Damage caused by cattle to different tree species recommended for silvipastoral systems. Pesquisa Florestal Brasileira, v. 32, n. 70, p. 183-92, Apr./Jun. 2012.
PROVENZA, F.D.; VILLALBA, J,J.; DZIBA, L.E.; ATWOOD, S.B.; BANNER, R.E. Linking herbivore experience, varied diets, and plant biochemical diversity. Small Rumin Res 49:257-274. 2003.
RALPHS, M.H.; PROVENZA, F.D. Conditioned food aversions: principles and practices, with special reference to social facilitation. Proc Nutr Soc 56:813-820. 1999.
ROMERO, C. Trees Responses to Stem Damage. Dissertation, University of Florida, 2006.
SAINT-ANDRIEUX, C.; BONENFANT, C.; TOIGO, C.; BASILLE, M.; KLEIN, F. Factors affecting beech *tica bark stripping by red deer us elaphus in a mixed forest*. Wildlife Biology. 15:187-196. doi: 10.2981/07100. 2009.
SILVA, F. de A. S. & AZEVEDO, C. A. V. de. A New Version of the Assistat-Statistical Assistance Software. In: WORLD CONGRESS ON COMPUTERS IN AGRICULTURE, 4, Orlando-FL-USA: Proceedings... Orlando: American Society of Agricultural and Biological Engineers, p.393-396. 2006.
SILVA, F. de A. S. & AZEVEDO, C. A. V. de. Principal Components Analysis in the Software Assistat-Statistical Attendance. In: WORLD CONGRESS ON COMPUTERS IN AGRICULTURE, 7, Reno-NV-USA: American Society of Agricultural and Biological Engineers, 2009.
SILVA, F. de A. S. The ASSISTAT Software: statistical assistance. In: INTERNATIONAL CONFERENCE ON COMPUTERS IN AGRICULTURE, 6, Cancun, Anais... Cancun: American Society of Agricultural Engineers, 1996. p. 294-298. 1996.
SUGAMOSTO, M. L. Use of geoprocessing techniques to draw up the agricultural suitability map and assess the suitability of use of the Cangüiri experimental station center, municipality of Pinhais - Paraná. 149f. Dissertation (Master's Degree in Agronomy) - Postgraduate Program in Agronomy, Federal University of Paraná. 2002.
VILLALBA, J. J.; MILLER, J.; HALL, J. O. et al. Preference for tanniferous (*Onobrychis viciifolia*) and non-tanniferous (*Astragalus cicer*) forage plants by sheep in response to challenge infection with *Haemonchus contortus*. Small Ruminant Research 112:199-207, 2013.
VILLALBA, J. J.; MILLER, J.; UNGAR, E. D. et al. Ruminant selfmedication against gastrointestinal nematodes: evidence, mechanism, and origins II. Parasite, http://dx.doi.org/10.1051/parasite/2014032, 2014.
WEMELSFELDER, Francoise. Animal boredom: is a scientific study of the subjective experiences of animals possible? In: Advances in Animal Welfare Science 1984. Springer Netherlands, p. 115-154. 1985.

WESTOBY, M. An analysis of diet selection by large generalist herbivores. Am Nat 108:290-304. 1974.
ARNOLD, R.J.; CLARKE, B. LUO, J. Trial of cold-tolerant eucalypt species in cooler regions of South Central China. Canberra: ACIAR, 106 p. (ACIAR Technical reports, 57). 2004.
BENIN, C. C.; WIONZEK, F. B.; WATZLAWICK, L. F. Initial assessments on the plantation of *Eucalyptus benthamii* Maiden et Cambage deployed in different spacing. Applied Research & Agrotechnology, Guarapuava-PR, v.7, n.1, p.55-61, 2014.
BENIN, Cristiane Carla; WIONZEK, Francielle Brandalise; WATZLAVICK, Luciano Farinha. ANNUAL INCREASE IN DIAMETER AND HEIGHT IN A PLANTATION OF Eucalyptus benthamii Maiden et Cambage UNDER DIFFERENT SPACING. Proceedings of the 4th Paranaense Forestry Congress, 2012. Available http://malinovski.com.br/CongressoFlorestal/Trabalhos/04-Management_Plantations/MFPlantations-Article-13.pdf, accessed 26/10/2016
BERNARDI, C. A. Evaluation of different soil preparation systems on the initial development of *Eucalyptus benthamii* Maiden et Cambage in the region of Guarapuava, PR. Irati, 46p. Dissertation (Master's Degree in Forestry Sciences) - Midwestern State University, 2010.
BERNARDI, Camila Motta Marin; MACEDO, H. R., PINHEIRO, R. S. B., & FREITAS, M. L. M. Eucalyptus planted forests in silvipastoral systems and the impact of animal input. Revista Verde de Agroecologia e Desenvolvimento Sustentável, v. 9, n. 5, p. 125-132, dec, 2014.
BERNARDINO, F. S. and GARCIA, B. Silvipastoral systems. Pesquisa Florestal Brasileira, Colombo, n.60, p.77-87, dec. Special Issue. 2009.
BERNARDO, A.L. Growth, biomass production and nutritional efficiency of three species of *Eucalyptus spp.* under different population densities in the cerrado region of Minas Gerais. Master's thesis. Federal University of Viçosa. Viçosa, 88p. 1995.
BOTELHO, S. A. Spacing. *In*: Scolforo. J. R. S. Manejo Florestal. Lavras: UFLA/FAEPRE, P. 381-405. 1998.
CALDEIRA, M.V.W.; SCHUMACHER, M.V.; SCHEEREN, L.W.; WATZLAWICK, L. F. O. Kuntze. Hypsometric relationship for *Araucaria angustifolia* (Bert.) in the western region of the state of Paraná. Revista Acadêmica: ciências agrárias e ambientais, Curitiba, v.1 n.2, p.79-88, 2003.
CALIL, F. N.; VIERA, M.; SCHUMACHER, M. V.; LOPES, V. G.; WITSCHORECK, R. Biomass and nutrients in an agroforestry system in the extreme south of Brazil. Ecología e Nutriçao Florestal, Santa Maria-RS, v.1, n.2, p.80-88, mai./ago., 2013.
CARVALHO, M. M.; ALVIM, M. J.; XAVIER, D. F. et al. Establishment of silvipastoral systems: emphasis on mountainous areas and low-fertility soils. Juiz de Fora: Embrapa Gado de Leite (Technical Circular, 68). 2002.
CARVALHO, M. M.; FREITAS, V. P.; ANDRADE, A.C. Initial growth of five tropical grasses in an angico-vermelho (*Anadenanthera macrocarpa* Benth.) understory. Past. Trop., v.17, p.24-30, 1995.
CARVALHO, Margarida M.; XAVIER, D. F. Sistemas silvipastoris para recuperaçao e desenvolvimento de pastagens. Juiz de Fora: Embrapa Gado de Leite-FAO, 2000.
DIAS-FILHO, M. B.; FERREIRA, J. N. Barriers to the adoption of silvipastoral systems. In: SIMPÓSIO DE FORRAGICULTURA E PASTAGENS, Lavras. Hot topic: cost-benefit ratio: proceedings. Lavras: NEFOR: UFLA, p. 347-365. 2007.
DIAS, L. P. R.; GATIBONI, L. C.; ERNANI, P. R.; MIQUELLUTI, D, J.; CHAVES, D. M.; BRUNETTO, G. Partial replacement of soluble phosphate with natural phosphate in the planting of *Eucalyptus benthamii* and *Eucalyptus dunnii* in the Southern Plateau of Santa Catarina. Ciência do Solo, Viçosa, v.38, n.2, p.516-523, 2014.
DIAS, Luciana Patrícia Rosa. Phosphorus and boron in the fertilization of *Eucalyptus dunnii* and *Eucalyptus benthamii* in soils of the Planalto Sul Catarinense. 2016. 158f. Thesis (PhD in Soil Science - Area: Plant Fertility and Nutrition) - Santa Catarina State University. Postgraduate Program in Agricultural Sciences, Lages, 2016.
FOTAN, I. C. I. Canopy dynamics and growth of eucalyptus clones subjected to pruning in an agroforestry system. 2007. 68 f. Dissertation (Master's Degree in Forest Science) - Federal University of Viçosa, Viçosa, 2007.
GARCIA, R.; COUTO, L.; ANDRADE, C. M. S.; TSUKAMOTO FILHO, A. de A. Silvipastoral systems in the southeast: CMM's experience. Text of the lecture presented at the

Seminar "Agroforestry Systems and Sustainable Development" - Campo Grande MS, 2003. http://saf.cnpgc.embrapa.br/publicacoes/22.pdf. Accessed on: 09 Nov. 2016.
HIGA, A. R.; CARVALHO, P. E. R. Survival and growth of twelve eucalyptus species in Dois Vizinhos, Paraná. Silvicultura, Sao Paulo, v. 2, n. 42, p. 459-461, 1990.
HIGA, R. C. V.; PEREIRA, J. C. D. Potential Uses of *Eucalyptus benthamii* Maiden et Cambage. Colombo: Embrapa Florestas, 4p. (Embrapa Florestas. Comunicado Técnico, 100). 2003.
IBÁ (BRAZILIAN TREE INDUSTRY). Ibá 2014. Brasília: 100 p., 2014.
JOSE. F. Agroforestry for ecosystem service and environmental benefits: an overview. Agroforestry Systems. v. 76, p. 1-10, 2009.
KLEINPAUL, I. S.; SCHUMACHER, M. V.; VIERA, M.; NAVROSKI, M. C. Mixed planting of *Eucalyptus urograndis* and *Acacia mearnsii* in an agroforestry system: I - Biomass production. Ciência Florestal, Santa Maria, v. 20, n. 4, p. 621-627, Oct.-Dec., 2010.
LAURA, Valdemir Antônio; ALVES, Fabiana Villa, ALMEIDA, Roberto Giolo de / technical editors. Agroforestry systems: sustainable agriculture. -- Brasília: Embrapa. 208 p., 2015.
LEITE, F. P.; BARROS, N. F. de; NOVAIS, R. F. de; SANS, L.A; FABRES, A. S. Growth of *Eucalyptus grandis* at different population densities. Revista Árvore, Viçosa, MG; v. 21, n. 3, p. 313-321, jul/ser. 1997.
LUNELLI, N. P.; RAMOS, F. S.; OLIVEIRA JÚNIOR, C. J. F. Agroforestry and externalities. Revista Verde (Mossoró - RN - BRASIL), v. 8, n. 5, p. 163 - 170, (Special Issue) December, 2013.
MACEDO, R. L. G., VALE, A.B., VENTURIN, N. Eucalyptus in agroforestry systems. Lavras: UFLA, 331p., 2010.
MACEDO, R. L. G.; BEZERRA, R. G.; VENTURIN, N.; VALE, R. S.; OLIVEIRA, T. K. Silvicultural performance of eucalyptus clones and agronomic characteristics of corn grown in a silvicultural system. Revista Árvore, Viçosa-MG, v.30, n.5, p.701-709, 2006.
MOTT, G. O.; LUCAS, H. L. The design, conduct and interpretation of grazing trials on cultivated and improved pastures. In: INTERNATIONAL GRASSLAND CONGRESS, 6, 1952, Pennsylvania. Proceedings..., Pennsylvania, p. 1380 - 1385. 1952.
NAIR, P. K. R. Perspective: The coming of age of agroforestry. J Sci Food Agric 87:1613-1619. 2007.
NISGOSKI, S.; MUÑIZ, G. I. B.; KLOCK, U. Anatomical characterization of the wood of *Eucalyptus benthamii* Maiden et Cambage. Ciência Florestal, Santa Maria, v.8, n.1, p. 67-76 67. 1998.
OECD/FAO (*ORGANIZATION FOR ECONOMIC COOPERATION AND DEVELOPMENT / FOOD AND AGRICULTURE ORGANIZATION OF THE UNITED NATIONS*). OECD-FAO Agricultural Outlook 2012-2021. OECD Publishing and FAO. 2012.
OLIVEIRA NETO, S. N., PAIVA, H.N. Implementation and management of the tree component in an agroforestry system. In: Oliveira Neto, S. N., Vale, A.B., Nacif, A.P., Vilar, M.B., Assis, J.B. (Ed.). Agrosilvipastoral system: crop, livestock and forest integration. Viçosa:
Sociedade de Investigaçoes Florestais, p.15-68, 2010.
OLIVEIRA NETO, S. N.; REIS, G. G.; REIS, M. G. F.; NEVES, J. C. L. Biomass production and distribution in Eucalyptus camaldulensis Dehn. in response to fertilization and spacing. Revista Árvore, Viçosa, v. 27, n. 1, p. 15-23, 2003.
OLIVEIRA, T. K. Agroforestry system with eucalyptus and brachiaria under different structural arrangements in a cerrado area. 150p. Thesis - PhD in Forestry Engineering - Federal University of Lavras - MG. 2005.
OLIVEIRA, T. K., MACEDO, R. L. G., VENTURIN, N., & HIGASHIKAWA, E. M. Silvicultural and productive performance of eucalyptus under different spatial arrangements in an agroforestry system. Pesquisa Florestal Brasileira, v. 1, p. 1-9, 2009.
PALUDZYSZYN FILHO, E. P.; SANTOS, P. E. T. Embrapa Florestas' eucalyptus genetic improvement program: results and prospects Colombo: Embrapa Florestas, 2011.
PALUDZYSZYN FILHO, E. P.; SANTOS, P. E. T.; FERREIRA, C. A. Eucalyptus Trees Suitable for Planting in the State of Paraná. Colombo: Embrapa Florestas, 2006.
PEREIRA, Ana Carolina Machado, DE ALMEIDA, J. C. C., MOREIRA, T. G. B., ZANELLA, P. G., DE CARVALHO, C. A. B., DE MORAIS, L. F. & Lima, M. A. Evaluation of the Tree

and Forage Component of Silvipastoral Systems in the "Campos das Vertentes" Mesoregion of Minas Gerais. Revista Brasileira de Agropecuária Sustentável, v. 5, n. 1, 2015.
PEREIRA, J. C. D.; STURION, J. A.; HIGA, A. R.; HIGA, R. C. V.; SHIMIZU, J. Characteristics of the wood of some eucalyptus species planted in Brazil. Colombo: Embrapa Florestas, 113p. (Embrapa Florestas. Documentos, 38). 2000.
PORFIRIO-DA-SILVA, V., MORAES, A.; MOLETTA, J.L.; PONTES, L.S.; OLIVEIRA EB, PELISSARI A, de CARVALHO PCF. Cattle damage to different tree species recommended for silvopastoral systems. Pesquisa Florestal Brasileira 32:67-76. 2012.
PORFÍRIO-DA-SILVA, V., MEDRADO, M.J.S., NICODEMO, M.L.F., DERETI, R.M. Afforestation of pastures with timber forest species: implementation and management. Colombo: Embrapa Florestas, 48p, 2010.
QUINKENSTEIN, A.; WÖLLECKE, J.; BÖHM, C.; GRÜNEWALD, H.; FREESE, D.; SCHNEIDER, B. U.; HÜTTL, R. Ecological benefits of the alley cropping agroforestry system in sensitive regions of Europe. Environmental Science and Policy, v.12. p.1112-1121. 2009.
SARTÓRIO, Ian Pereira. Evaluation and modeling of the growth of eucalyptus energy forests planted at different densities. 2014. 136 f. Dissertation (master's degree) - Federal University of Paraná, Agricultural Sciences Sector, Postgraduate Program in Forestry Engineering. 2014.
SCOLFORO, J.R.S. Biometria Florestal 2, Técnicas de regressão aplicada para estimar: volume, biomassa, relação hipsométrica e múltiplos produtos de madeira. Lavras, UFLA/FAEPE/DCF, 292p., 1997.
SERPE, Edson Luis. Effects of different fertilizer dosages on the initial growth of *Eucalyptus benthamii* in the southern region of the state of Paraná -- Irati, PR. Dissertation (master's degree) - UNICENTRO, Postgraduate Program in Forestry Sciences, area of concentration in Forest Management. 82f. 2015.
SILVA, Luciana Duque. Genetic Improvement of *Eucalyptus benthamii* Maiden et Cambage for the Production of Sawn Wood in Areas where Severe Frosts Occur. Thesis (PhD) - Federal University of Paraná, Agrarian Sciences Sector, Postgraduate Program in Forestry Engineering. 2008.
SILVA, I. C. Agroforestry systems: concepts and methods. 1 ed. - Itabuna: SBSAF. 308 p., 2013.
SILVA, F. de A. S. e. & AZEVEDO, C. A. V. de. A New Version of the Assistat-Statistical Assistance Software. In: World Congress on Computers in Agriculture, 4, Orlando-FL-USA: Proceedings... Orlando: American Society of Agricultural and Biological Engineers, p.393-396. 2006.
SILVA, F. de A. S. e. & AZEVEDO, C. A. V. de. Principal Components Analysis in the Software Assistat-Statistical Attendance. In: World Congress on Computers in Agriculture, 7, Reno-NV-Usa: American Society of Agricultural and Biological Engineers, 2009.
SOUZA, K. K. F.; Effect of mineral fertilization on the growth and essential oil production of forest species in the first plateau of Paraná, Pinhais-PR. Thesis (PhD) - Federal University of Paraná, Agrarian Sciences Sector, Postgraduate Program in Forest Engineering, 142 f., 2015.
VIEIRA, M.; SCHUMACHER, M. V. Biomass in monospecific and mixed stands of eucalyptus and black wattle and corn in an agroforestry system. Cerne, Lavras, v. 17, n. 2, p 259-265, Apr./Jun. 2011.

More
Books!

I want morebooks!

Buy your books fast and straightforward online - at one of world's fastest growing online book stores! Environmentally sound due to Print-on-Demand technologies.

Buy your books online at
www.morebooks.shop

Kaufen Sie Ihre Bücher schnell und unkompliziert online – auf einer der am schnellsten wachsenden Buchhandelsplattformen weltweit! Dank Print-On-Demand umwelt- und ressourcenschonend produzi ert.

Bücher schneller online kaufen
www.morebooks.shop

info@omniscriptum.com
www.omniscriptum.com

OMNIScriptum

MIX
Papier aus verantwortungsvollen Quellen
Paper from responsible sources
FSC® C105338

Printed by Books on Demand GmbH, Norderstedt / Germany